D1189865

Advanced Textbooks in Control and Signal Processing

Springer
London
Berlin
Heidelberg
New York
Hong Kong
Milan
Paris
Tokyo

M.O. Tokhi, M.A. Hossain and M.H. Shaheed

Parallel Computing for Real-time Signal Processing and Control

Springer

M.O. Tokhi, PhD, CEng
Department of Automatic Control and Systems Engineering,
The University of Sheffield, Mappin Street, Sheffield, S1 3JD, UK

M.A. Hossain, MSc, PhD
Division of Electronics and IT, School of Engineering, Sheffield Hallam
University, Howard Street, Sheffield, S1 1WB, UK

M.H. Shaheed, MSc, PhD
Department of Engineering, Queen Mary, University of London, Mile End
Road, London, E1 4NS, UK

British Library Cataloguing in Publication Data
Tokhi, M.O.
    Parallel computing for real-time signal processing and
    control. – (Advanced textbooks in control and signal
    processing)
    1.Signal processing 2.Parallel processing (Electronic
    computers) 3.Parallel programming (Computer science)
    4.Real-time control
    I.Title II.Hossain, M.A. III.Shaheed, M.H.
    621.3'822
    ISBN 1852335998

Library of Congress Cataloging-in-Publication Data
A catalog record for this book is available from the Library of Congress.

ISSN 1439-2232
ISBN 1-85233-599-8 Springer-Verlag London Berlin Heidelberg
a member of BertelsmannSpringer Science+Business Media GmbH
http://www.springer.co.uk

Typesetting: Electronic text files prepared by authors
Printed and bound in the United States of America
69/3830-543210 Printed on acid-free paper SPIN 10868230

51559042

Series Editors' Foreword

The topics of control engineering and signal processing continue to flourish and develop. In common with general scientific investigation, new ideas, concepts and interpretations emerge quite spontaneously and these are then discussed, used, discarded or subsumed into the prevailing subject paradigm. Sometimes these innovative concepts coalesce into a new sub-discipline within the broad subject tapestry of control and signal processing. This preliminary battle between old and new usually takes place at conferences, through the Internet and in the journals of the discipline. After a little more maturity has been acquired by the new concepts then archival publication as a scientific or engineering monograph may occur.

A new concept in control and signal processing is known to have arrived when sufficient material has evolved for the topic to be taught as a specialised tutorial workshop or as a course to undergraduate, graduate or industrial engineers. *Advanced Textbooks in Control and Signal Processing* is designed as a vehicle for the systematic presentation of course material for both popular and innovative topics in the discipline. It is hoped that prospective authors will welcome the opportunity to publish a structured and systematic presentation of some of the newer emerging control and signal processing technologies.

Parallel computing brings together computing science and applications in both control and signal processing. The aim is fast, efficient, accurate real-time computing that can be used in time-critical algorithms that might be needed in such diverse fields as robotics, aerospace control systems and supply chain management in commerce. Computing using parallel processors is thus an implementational technology which enables the use of advanced control and data processing algorithms in demanding technological applications.

In this textbook, Drs. Tokhi, Hossain and Shaheed present a systematic introduction to the parallel computing field with special emphasis on its applications to control and signal processing. After a broad introductory chapter, the authors have chapters on parallel architectures, performance metrics, parallel programming and algorithms. The last two chapters of the book report on the hardware features of parallel computers as available to users and finally examine a set of typical control and signal processing applications. Performance metrics are given for the use of different parallel computing environments to run the various algorithms developed earlier in the textbook. This comprehensive textbook development of parallel computing with applications in control and signal processing can be used either as an advanced course textbook, a self-learning text or even a reference text for many of the specialist terms used in parallel computing. As such it is a versatile addition to the *Advanced Textbooks in Control and Signal Processing* series.

M.J. Grimble and M.A. Johnson
Industrial Control Centre
Glasgow, Scotland, U.K.
October, 2002

Preface

The computing performance demands in modern real-time signal processing and control applications are increasing at a rapid pace. These impose hard limits on requirements of computational capabilities and processing speed, which are most often not met by traditional computing methods. Parallel processing offers the potential for solving problems of this nature by devising suitable parallel computing methods. One of the main issues in this process is the partitioning of an application into tasks and the mapping of these tasks onto the processing elements (PEs). It is often noticed that the resulting parallel architecture does not offer the desired performance due to a mismatch between the computational requirements of the tasks and the computing capabilities of the PEs. This book aims at presenting a principled introduction to the design and development of real-time parallel architectures and computing methods on the basis of the interrelation between algorithms and architectures. This involves an exploration of the nature and computing requirements of typical algorithms commonly encountered in signal processing and control applications and an investigation into the nature and computational capabilities of sequential and parallel high-performance processors. The strategy adopted thus allows identification and, in turn, exploitation of the computational capabilities of processors for suitable task-to-processor matching. In addition to worked examples and end of chapter exercises, the book provides case studies demonstrating theoretical concepts within a practical setting and framework.

The material presented in this book has largely been derived from the research work carried out by the authors over several years. Accordingly, there have been several other colleagues and students involved in this process. These have included Professor Peter J. Fleming and Dr Daniela N. Ramos-Hernandez (University of Sheffield, UK), Dr Abul K. M. Azad (Northern Illinois University, USA), Michael J. Baxter (University of Wales Bangor, UK), Margarida M. Moura and Dr Graca Ruano (University of Algarve, Portugal), Professor Gurvinder S. Virk (University of Leeds, UK), Benjamin Chan (Sheffield Hallam University, UK), Upama Kabir (University of Dhaka, Bangladesh). The authors are indebted to the support and encouragement of their families; their patience and understanding during this project have been crucial to its successful completion. The authors would like also to acknowledge the enthusiastic encouragement and support of Professor Michael

Johnson (University of Strathclyde, UK). Furthermore, many thanks to the staff of Springer-Verlag (London) Ltd for their encouragement, advice, and patience.

M. O. Tokhi (Sheffield, UK)
M. A. Hossain (Sheffield, UK)
M. H. Shaheed (London, UK)
September 2002

Table of Contents

1. Introduction

1.1 Objectives

- To develop a conceptual understanding of parallel processing.
- To describe the requirements and benefits of parallel processing.
- To describe influential factors and the extent to which they require consideration to make parallel processing feasible.
- To demonstrate the importance of parallel processing through a description of selected practical applications.

1.2 Parallel Processing: Concepts and Evolution

In day-to-day life, people are required to complete many tasks within a limited timespan. Speed is the main issue of concern with such time-bound problems. Moreover, performance, efficiency and the smooth running of any system are highly dependent on speed. Of the mechanisms that are generally adopted to increase performance speed, parallel processing is most worthy of mention. It is utilised in many real-life systems. For example, consider the book-lending process in a university library, where students are served simultaneously by a number of librarians. With only one librarian, all students must form a single queue to be served. In that case, students will be in the queue for a longer time. If there are a number of librarians to serve the students, it will speed up the process significantly. Similarly, in a bank with only one cashier present, all customers must form a single queue to be served. With two cashiers, the tasks are effectively divided between them and customers are served twice as fast.

The basic concept of parallel processing is quite clear from the above two examples. It can be simply defined as completing a large task quicker and more efficiently by dividing it into several smaller tasks and executing them simultaneously using more resources. However, while applying parallel processing to a process, a number of factors are required to be considered, for example, whether the task in question can be performed in parallel, its cost effectiveness, synchronisation of the subtasks, and communication among the resources.

Practically, tasks that can effectively be divided are good candidates for parallel processing. Consider the first example above. If all book loans are required to be approved by the main librarian, the flow of book loans will not increase with an increase in the number of librarians as all book-lending students will be required to be in a single queue for loan approval by the main librarian.

Speed is probably the most frequently addressed term in computing. The very purpose of the invention and evolution of the microprocessor is to add speed to the normal flow of human civilisation. Like all other cases, speed is the main issue of concern in the computational world, mainly due to real-time requirements. Many scientific and engineering problems are required to be completed within specific time periods to produce accurate and valid results. They often involve huge repetitive calculations on large amounts of data. Some examples are: weather forecasting, the evolution of galaxies, the atomic structure of materials, the efficiency of combustion in an engine, damage due to impact, modelling DNA structure, the behaviour of microscopic electronic devices, controlling a robotic arm (Culler *et al.*, 1999). Some of the problems mentioned, such as weather forecasting, are real-time. There is no use forecasting weather for 10 am at 11 am on the same day. In the same way, the use of a robot is meaningless if the robotic arm reaches a certain position in 6 minute intervals whereas it is required to reach there in 5 minute intervals. Some other problems like modelling fluid dynamics are real-time in the sense that they have to be completed within a reasonable time to ensure the effectiveness and efficiency of the system. Besides the problems mentioned above, many recent applications, such as virtual reality, require considerable computational speed to achieve results in real-time (Culler *et al.*, 1999).

High computational speed of a processor is not only demanded by a problem, it is also the human tendency to get things done as fast as possible. Although today's computers are executing tasks with speeds as high as 2.2 GHz; this is not something that people are satisfied with, just as they have not been previously. On the contrary, algorithms in almost all application domains are becoming more and more complex to meet modern automation and precision demands. It seems that, there will be no end to this race. Thus, regardless of the computational speed of current processors, there will always be applications that will require more computational power.

It is reasonable to pose the question: how can the speed of a processor be increased to a desired level? A considerable amount of effort has already been devoted to making computers perform faster and faster; improving hardware design circuitry, instruction sets and algorithms, to name a few. However, there are physical limitations to all these improvements. Algorithms and instruction sets cannot be improved beyond a certain limit. The density of transistors within a processor limits improvement to the design of circuitry. A certain density limit cannot be exceeded as the transistors create electromagnetic interference for one another beyond this limit. However, this is not the end. More importantly, processor speed depends on the speed of transmission of information between the electronic components within the processor, which is, in fact, limited by the speed of light. The speed of light is 30 cm per ns ($1 s = 10^9$ ns) and this speed can be achieved only by means of optical communication, which is yet to be realised. Even this speed is not sufficient for some users to execute their applications.

Measures have also been taken to use specialised high-performance processing elements (PEs) to meet the computational speed demand. For example, digital signal processing (DSP) devices are designed in hardware to perform concurrent add and multiply instructions and execute irregular algorithms efficiently. Vector processors are designed to process regular algorithms involving matrix manipulation efficiently. The ideal performance of these processors demands a perfect match between the capability of their architecture and the program behaviour. These include algorithm design, data structures, language efficiency, programmer's skill and compiler technology (Ching and Wu, 1989; Hwang, 1993; Tokhi *et al.*, 1997a,b, 1999a,b). However, in many cases purpose-built PEs are not efficient enough to bridge the gap between the hardware and the speed requirements of applications.

In light of the problems mentioned above, parallel processing has emerged as a potential solution to speed requirement. The concept of parallel processing was described in terms of real-life problems above. With reference to computing, it corresponds to the execution of a single program faster by dividing it into several tasks and allocating these tasks to processors to execute simultaneously without affecting the final result. The approach is expected to be very effective at increasing the speed of performance of program execution. In an ideal situation, a parallel architecture comprising n processors will provide n times the computational speed of a single processor and the application would be completed in $1/n$ th time of a single processor implementation. For example, consider Figures 1.1 and 1.2. In Figure 1.1, a processor takes two seconds to execute a program whereas if that program is divided into two subprograms and executed simultaneously with two processors, it takes only one second, which is half the execution time of one processor. However, this ideal situation is rarely achieved in practice. Very often applications cannot be perfectly divided into equal independent components and allocated to processors. Furthermore, communication among the processors and synchronisation of the parts are inevitable to obtain perfect results. However, significant improvement may be achieved, depending upon the nature of the application in question, such as the amount of parallelism present in it.

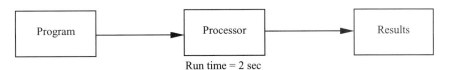

Run time = 2 sec

Figure 1.1. Sequential processing

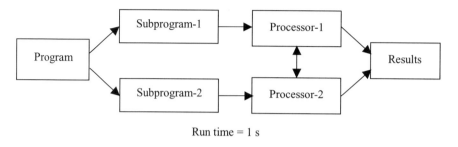

Run time = 1 s

Figure 1.2. Parallel processing

The idea of parallel processing described above is well established. It is evident from the literature that parallel computing machines were discussed at least as far back as the 1920s (Denining, 1986). It is noted that, throughout the years, there has been a continuing research effort to understand and develop parallel computing (Hocney and Jesshope, 1981). As a result various forms of parallel processing and parallel architectures have evolved. Some forms of parallel processing worth noting are shared memory parallel processing, message passing, and data parallel processing. In the case of shared memory parallel processing, communication among the tasks is established through some shared locations. Small-scale shared memory multiprocessors came into being with the inception of mainframe computers. However, the extensive use of shared memory parallel processing started in the mid-1980s (Culler *et al.*, 1999). Communication and synchronisation among the processes is established through well-controlled send and receive mechanisms. The origins of message passing machines emerged in the RW4000, introduced in 1960 (Culler *et al.*, 1999). This type of parallel processing is quite commonly used today. In the case of data parallel processing, operations are performed in parallel on each element of a large regular data structure such as an array or matrix. Data parallel machines also emerged with the design of the Solomon computer (Culler *et al.*, 1999). Flynn (Flynn, 1972) developed taxonomy of such type parallel computers in the early 1970s. Over the years, tremendous innovation with massive diversity has occurred in the development of parallel processing, which this book intends to address in a coherent manner.

1.3 Real-time Systems and Parallel Processing

There are a number of definitions of real-time systems in the literature, such as:

Real-time systems are those which must produce correct responses within a definite time-limit. Should computer responses exceed these time bounds then performance degradation and/or malfunctions results

(Bennett, 1988)

Real-time systems are defined as those systems in which the correctness of the system depends not only on the logical result of computation, but also on the time in which the results are produced

(Stankovic, 1997)

There are varieties of real-time systems present around us, for example air traffic control systems, process control systems, flight control systems, space systems, intensive care monitoring, intelligent highway systems, mobile and wireless computing, and multimedia and high-speed communication systems. Recently, some of these real-time systems have included expert systems and artificial intelligence methods, creating additional requirements and complexities.

In terms of time limits, a real-time system can be a *hard real-time system* where there is no value in executing tasks after their deadlines have passed, or a *soft real-time system* where tasks retain some diminished value after their deadlines. In this case these tasks should still be executed, even if they miss their deadlines.

The time constraints for tasks in a real-time system could be periodic, aperiodic or sporadic. A periodic task is one that is activated once every T units of time. The deadline for each activated instance may be less than, equal to, or greater than the period T. An aperiodic task is activated at unpredictable times. A sporadic task is an aperiodic task with an additional constraint that there is a minimum interarrival time between task activations. In a real-time system, the characteristics of the various application tasks are usually known *a priori* and might be scheduled statically or dynamically. A static specification of schedules is used for periodic tasks, whereas for aperiodic tasks a dynamic specification is used. A real-time system that is designed in a static manner is not flexible and may incur lower run-time overheads. In practice, most applications involve a number of components that can be statically specified along with many dynamic components.

The characteristics of a real-time system (Stankovic, 1997) include the following:

Granularity of the deadline and laxity of the tasks. In a real-time system some of the tasks have deadlines and/or periodic timing constraints. If the time between when a task is activated and when it must complete execution is short, then the deadline is tight. This implies that the operating system reaction time has to be short, and the scheduling algorithm to be executed must be fast and very simple. Tight time constraints may also arise when the deadline granularity is large, but the amount of computation required is also great. In other words even large granularity deadlines can be tight when the laxity (deadline minus computation time) is small.

Strictness of deadline. This refers to the value of executing a task after its deadline. Whereas for a hard real-time task there is no value in executing the task after the deadline has passed, for a soft real-time task some diminished value is returned after its deadline, and therefore it should still be executed.

Reliability. The requirement for critical tasks should be such that they always meet their deadline (a 100% guarantee), subject to certain failure and workload assumptions.

Size of system and degree of coordination. Real-time systems vary considerably in size and complexity. The ability to load entire systems into memory and to limit task interactions simplifies many aspects of building and analysing real-time systems.

Environment. The environment in which a real-time system is to operate plays an important role in the design of the system. When the environment is defined or deterministic the deadlines are guaranteed *a priori*. However, a system may be large, complex, distributed, adaptive, containing many types of timing constraints and need to operate in a highly non-deterministic environment.

In the area of hard real-time systems many important developments have been made in real-time scheduling, operating systems, architecture and fault tolerance, communication protocols, specification and design tools, formal verification, databases and object-oriented systems. In contrast, distributed multimedia has produced a new set of soft real-time requirements (Ramos-Hernandez, 1998).

One of the main reasons for the rapid evolution of parallel processing is due to the timing requirements of real-time systems. It is worthless in the case of hard real-time systems if the system produces results after the deadline has passed. In the case of soft real-time systems parallel processing may help to produce results within the expected time limit to achieve better performance. In fact, algorithms and related aspects of modern real-time applications are becoming so complex that it is almost impossible in many cases to meet the deadline using only general-purpose or special-purpose processors and parallel processing is a possible solution in those cases.

1.4 Basic Components of Parallel Processing

As mentioned above, parallelism enables multiple processors to work simultaneously on several parts of a program in order to execute it faster than could be done otherwise. As such, parallel processing entails some basic components for implementation.

- Computer hardware that supports parallelism. This hardware is designed to work with multiple processors and may all reside either on the same computer or may be spread across several computers. When spread across several computers, each computer is referred to as a node. A means of communication among the processors or nodes is required for message transmission for synchronisation of various parts of the program to ensure correctness of the final result.
- Operating system and application software that are capable of managing multiple processors.

1.5 Parallel Processing: Tasks, Processes, Processors and Basic Characteristics

To understand parallel processing at the implementation level, one must know some basic terms such as task, process, processors and the fundamental characteristics of parallel processing. These are described below.

1.5.1 Task

A task is the smallest unit of the work that needs to be done. In other words, it is the independently executable smallest unit of a program that is executed by one processor and concurrency among processors is exploited only across tasks. Tasks can be of two types, namely fine-grained and coarse-grained. If the amount of work a task performs is small it is called a fine-grained task and otherwise it is called a coarse-grained task.

1.5.2 Process

A process performs the tasks assigned to it. Usually more than one task is assigned to a process. However, a process can consist of only one task. A program in the parallel executable format is composed of a number of processes, each of which performs a subset of tasks in the program. Processes usually need to communicate and synchronise with one another to perform tasks assigned to them by executing them on the physical processors in the machine.

1.5.3 Processor

While the task and process are software-based entities, the processor is a hardware-based physical resource. A processor executes the process assigned to it. Usually a parallel program is written in terms of processors, and the number of processes in the program could be the same as the number of processors, or higher or less than the number of processors in the parallel architecture. If the number of processes is more than the number of processors, more than one process is assigned to some or all of the processors, depending on the number of processes and nature of the algorithm. If the number of processes is less than the number of processors, some processors are not assigned processes and they remain idle during execution of the program.

1.5.4 Basic Characteristics

A parallel processing system has the following characteristics:

- Each processor in a system can perform tasks concurrently. They might have their own memory units for executing and managing the tasks and/or can share a common memory unit.

- Tasks may need to be synchronised. Sometimes synchronisation is vital to ensure correctness of the results.
- Processors or nodes usually share resources, such as data, disks, and other devices.
- A parallel system executes tasks or programs faster than sequential processing of the same tasks or programs.

1.6 Levels of Parallelism

Parallelism is applicable at various processing levels mainly on the basis of computational grain size (granularity). There are five levels of parallelism, namely, job (program) level, subprogram level, procedure level, loop level, instruction (expression) level and bit level. These are shown in Figure 1.3.

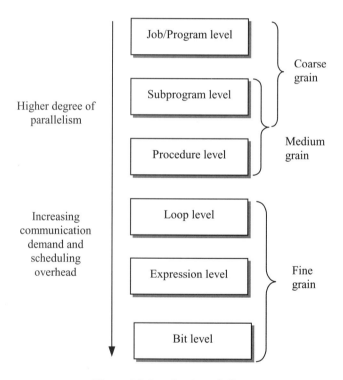

Figure 1.3. Levels of parallelism

Program or job level is the highest level of parallelism that occurs in any multiprogrammed system when multiple jobs/programs are executed concurrently using time-sharing. The grain size can be as high as tens of thousands of instructions in a single program.

Subprogram level is next to program level. At the level of subprograms procedures or subroutines can be executed in parallel. The grain size may typically contain thousands of instructions (medium- or coarse-grain). Multifunction processors and array processors can provide parallel execution at this level.

Procedure level corresponds to medium-grain size at the task, procedure, subroutine and co-routine levels. A typical grain contains less than 2000 instructions. Multitasking belongs to this category.

Loop level corresponds to iterative and non-recursive loop operations. A typical loop contains less than 500 instructions and thus this level is considered as fine-grained parallelism. Vector processing is mostly exploited at the loop level.

Expression level is fine-grained parallelism. In this level parallel processing occurs between phases of an individual instruction/statement. A typical grain contains less than 20 instructions.

The lowest level of parallelism is the bit level. It happens within an instruction and hence is fine-grained.

1.7 Parallel Processing: Steps Involved

Sequential algorithms that need to be parallelised, are normally given in the form of description, pseudocode, algorithm or program. Some of these algorithms are easily decomposable into tasks and processes and hence require less effort for parallelisation. But there are some algorithms that require considerable research for parallelisation and others are not at all parallelisable or cost effective for parallelisation. Whatever the nature of the algorithm the goal of parallelisation is to obtain high performance and increased speed over the best sequential program that solves the same problem. This is to ensure efficient load balancing among processors, reduction in communication overhead and synchronisation. Several steps are involved in parallelising an algorithm, each of which contributes to achieve better performance. The steps are:

- Decomposing computation into tasks.
- Assigning tasks to processes.
- Orchestration: data accessing and establishing communication and synchronisation among processes.
- Mapping processes to processors for execution.

1.7.1 Decomposition

At this very first stage, the sequential algorithm is analysed and examined to find the amount of parallelisation inherent in the algorithm and thus the total computation is decomposed into an appropriate number of tasks. The load could be fixed or variable. In the case of fixed load size, decomposition may sometimes be straightforward. However, in the case of varying load size decomposition could be quite critical since tasks become available dynamically as the program executes. Decomposition has a substantial impact on the performance of a parallel process. To ensure better performance all processes are required to be kept busy as much of the time as possible through efficient decomposition.

1.7.2 Assigning Tasks to Processes

In this step tasks are allocated among the processes with the objectives of

i.   balancing the workload among processes. This is often referred to as load balancing;
ii. reducing the interprocess communication and runtime overhead.

Load balancing mainly includes data access, computation and communication. Reducing interprocess communication is very important for achieving better performance especially when a number of processors are involved in the parallel execution of the algorithm. Allocation of tasks can be of two types: static allocation and dynamic allocation. In the case of static allocation tasks are allocated to processes by analysing the algorithm before execution has started. Allocation does not change during program execution. In the case of dynamic allocation tasks are allocated at runtime during program execution.

1.7.3 Orchestration

Accessing data, exchanging data among the processes as needed and ensuring appropriate synchronisation belong to this step. Programming plays a vital role in performing all these actions. The main factors needing to be considered in this step include organising data structures, scheduling the tasks assigned to processes, whether to communicate explicitly or implicitly and how exactly to organise and express the interprocess communication and synchronisation that resulted from the assignment (Culler *et al.*, 1999). The main objectives in this step include reducing the cost of communication and synchronisation, preserving locality of data reference and reducing the overhead of parallelism management.

1.7.4 Mapping Processes to Processors for Execution

This step involves mapping the processes to processors for execution. This can be performed manually by the programmer or with the help of an appropriate scheduling algorithm and operating system. The processes are not supposed to migrate from one processor to another during execution.

1.8 Parallel Processing: Advantages

Parallel processing offers several advantages over sequential processing. These are described below.

1.8.1 Speedup

Parallel processing reduces execution time of the program considered and hence increases speedup or improves response time. Speedup is defined as the ratio of the execution time with one processor to the execution time using a number of processors, and is calculated as:

$$Speedup = \frac{T_1}{T_n} \tag{1.1}$$

where, T_1 is the time taken by one processor to execute a task and T_n is the time taken by n processors to execute the same task. For example, consider Figures 1.1 and 1.2, where one processor takes 2 seconds to execute a particular program whereas a parallel configuration of two processors takes 1 second to execute the same program thus, in this case

$$Speedup = \frac{2}{1} = 2.0$$

Here two processors take half the time taken by a single processor to execute the program. This is an ideal situation, which is rarely achieved, in real life. The ideal speedup curve is shown in Figure 1.4, and is a line drawn at $45°$ angle to both axes.

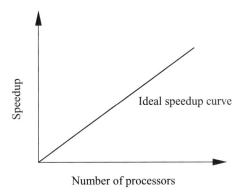

Figure 1.4. Ideal speedup curve

The ideal speedup curve is seldom achieved due to such factors as communication overhead and memory access. It is rarely possible to parallelise an algorithm without communication among the processors involved. On the other hand, some programs can easily be divided into parts and allocated to processors whereas others do not lend themselves to this mechanism.

1.8.2 Scaleup

Scaleup or high throughput of a system can be defined as the ability to maintain a constant execution time period as the workload or job size increases by adding additional hardware such as processors, memory and disks. While calculating the speedup, the problem size is kept fixed whereas scaleup is calculated by increasing the workload size or transaction volume. Scaleup is measured in terms of how much transaction volume can be increased by adding more hardware, specially processors, while still maintaining constant execution time. Scaleup is calculated by:

$$Scaleup = \frac{W_N}{W_1} \tag{1.2}$$

where W_1 is workload or transaction volume executed in a certain amount of time using one processor and W_N is the workload executed in the same time using n processors. For example, a single processor system supports a workload of 10 transactions per second, whereas a system with 5 processors can provide the same response time with a workload of 50 transactions, thus the scaleup would be

$$Scaleup = \frac{50}{10} = 5$$

This corresponds to an ideal situation.

Figure 1.5 shows the ideal scaleup curve where execution time remains constant by increasing the number of processors as the transaction volume is increased. In practice, about a certain point the execution time increases with increase of the workload even if additional processors are added.

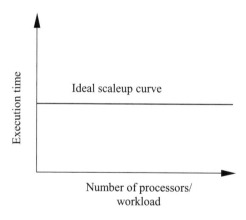

Figure 1.5. Ideal scaleup curve

Parallel systems provide much better scalability in comparison to single-processor systems as added workload or transaction volume over time can easily be handled by adding extra processing power (especially processors) without loss of response time.

1.8.3 Fault Tolerance

A parallel processing system is more likely to be fault tolerant than a single-processor system, as parallel systems offer the flexibility of adding and removing processors or other additional hardware in the system. For example, if a processor in a parallel system does not work for some reason it could easily be replaced with another one without affecting the ongoing work. This is not possible in the case of a single-processor system.

1.8.4 Cost-to-Performance Ratio

The design and development of a faster computer increases costs significantly. Accordingly, increasing the processing power on a single processor becomes technically difficult and very expensive beyond a certain limit. In such cases parallel systems provide better performance for the price.

1.8.5 Handling Larger Tasks

It is easily understood that a parallel system can handle bigger problems than a single-processor system. If the problem tends to grow over time, parallel processing can cope with it by adding additional hardware.

1.9 Factors Influencing Performance

It is evident from the discussion above that parallel processing offers a number of advantages over sequential processing. However, these advantages are not easily attainable. There are various factors that influence the performance of parallel processing, as described below.

1.9.1 Amount of Inherent Parallelism Present in the Algorithm

The amount of parallelism present in the algorithm influences the performance of parallelisation significantly. All parts of an algorithm are, normally, not parallelisable. If 80 – 90% of an algorithm is absolutely sequential in nature, it might not be worth parallelising the algorithm. If the portion of the sequential task in an algorithm is η and no overhead is incurred then it follows from *Amdahl's law* (Amdahl, 1967), described in Chapter 4, that the maximum parallel speedup is limited by η. That is, with a parallel architecture of N processors:

$$\lim_{N \to \infty} S_N = \frac{1}{\eta} \qquad (1.3)$$

For example, with only 10% of the computation being sequential, the maximum speedup is 10.

1.9.2 Structuring Tasks

Structuring tasks for parallel processing is a major challenge. A large task needs to be decomposed into several smaller tasks and allocated among the processors to execute them simultaneously. Not all the tasks are decomposable and thus are not suitable for execution using a parallel system. However, many tasks are decomposable. A task might be decomposed and structured in many ways and to find the best decomposable structure is challenging. While decomposing a task, there are certain issues to consider such as how to keep the overheads to a minimum level and maintain synchronisation effectively.

1.9.3 Synchronisation

Synchronisation is another critical factor for successful parallel processing. Co-ordination of the concurrent tasks allocated among the processors is called synchronisation, and is necessary for the correctness of results. Tasks are required to be divided up and allocated to processors, keeping the requirement for synchronisation as low as possible to ensure better speedup and scaleup. Synchronisation depends upon a number of factors such as the nature of the task or algorithm, the amount of resources, and the number of users and tasks working with the resources. To make synchronisation cost effective is another challenge. Different kinds of parallel processing software may achieve synchronisation, but a given approach may or may not be cost effective. Sometimes synchronisation can be accomplished very cheaply and in other cases the cost of synchronisation might be very high.

1.9.4 Overhead

Overhead is a vital issue to consider when implementing parallel processing. It is always beneficial to keep the overhead as low as possible. The less the overhead the better the speedup and scaleup. Too much overhead may diminish the benefits of parallel processing. There are several forms of overhead that limit the speedup of parallel processing. The important ones include

- Time slots when all processors are not active, i.e., some of the processors are sitting idle.
- Additional computations that occur due to parallelising the algorithm such as recomputing the constants locally.
- Communication time during message passing.

Of these, the overhead due to communication dominates in most cases where message passing among the processors is a prerequisite for parallelisation. In fact parallel execution time is considered to be composed of two parts, the computation part and the communication part, and is given as:

$$T_p = T_{comp} + T_{comm} \tag{1.4}$$

where T_p is the parallel execution time, T_{comp} is the computation time and T_{comm} is the communication time. The communication time T_{comm} is expressed as:

$$T_{comm} = T_0 + \frac{n}{B} \tag{1.5}$$

where T_0 is the startup time, n is the amount of data and B is the data transfer rate, usually given in bytes per second. The communication time depends on a number of factors including the size of the message, the interconnection of the parallel structure and the mode of data transfer. If p messages each containing n data items need to be sent, the total communication time would be

$$T_p = n(T_{comp} + T_{comm})$$

Thus, the communication overhead has a considerable impact on the performance of a parallel architecture. The performance of a parallel architecture in terms of communication time and computation time is given by:

$$P = \frac{T_{comp}}{T_{comm}} \tag{1.6}$$

This implies that the more the communication time, the worse the performance.

1.9.5 Message Passing

Message passing is an inevitable requirement in the case of message passing based parallel processing. In message passing a substantial distance exists between two processors. Message passing is normally established through send and receive mechanisms between two processes. The send process specifies a local data buffer that is to be transmitted to the receive process and the receive process specifies a local data buffer into which data is to be placed. The send process sends a request signal to the receive process for message passing. After getting acknowledgement from the receive process the send process starts data transmission and the receive process starts to receive the transmitted data as shown in Figure 1.6. Synchronisation, communication and co-operation are inevitable between the send and receive processes. Fast and efficient messaging among the processors is required to ensure good performance of parallel processing. A system with high bandwidth is capable of transmitting more messages. Bandwidth is the total size of message that can be sent per second.

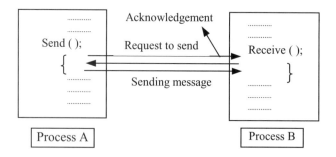

Figure 1.6. Message passing between two processes

1.10 Parallel Processing: Some Applications

Parallel processing has a wide range of applications in various fields that range from weather forecasting to robotics. Some of the major applications of parallel processing are described below.

1.10.1 Weather Forecasting

Weather forecasting is a real-life example of parallel processing. Satellites used for weather forecasting collect millions of bytes of data per second on the condition of earth's atmosphere, formation of clouds, wind intensity and direction, temperature, and so on. This huge amount of data is required to be processed by complex algorithms to arrive at a proper forecast. Thousands of iterations of computation may be needed to interpret this environmental data. Parallel computers are used to perform these computations in a timely manner so that a weather forecast can be generated early enough for it to be useful.

1.10.2 Motion of Astronomical Bodies

Predicting the motion of astronomical bodies requires a huge number of calculations. There are numerous bodies in space and these are attracted to each other by gravitational forces. Although the calculation is done using simple formulae, but a massive amount of calculation is required to predict the motion of each body as forces from many other bodies are acting on it. In fact, if there are N bodies in the space, there will be $N-1$ forces acting on each body, so there will be approximately N^2 calculations and these calculations must be repeated with each change of position. Parallel processing is required for such calculations.

1.10.3 Database Management

Parallel processing is widely used in database management systems. With the developments in and demand for precision, database sizes are increasing and queries are becoming more complex especially in data warehouse systems. Some databases referred to as very large databases hold several terabytes of data. Complex queries are run on these data warehouses to gather business intelligence and to aid decision-making. Such queries require substantial processing time to execute. By executing these queries in parallel the elapsed time can be reduced.

1.10.4 Satellite, Radar and Sonar Applications

Most satellite, radar and sonar applications handle vast amounts of data and varieties of signals. Parallel processing is used to process these data and signals. In fact, many of those applications cannot be even thought of without parallel processing.

1.10.5 Aerospace Applications

The computing requirements of safety-critical aerospace control systems are currently doubling every 2–3 years. This trend is widely expected to continue in future in response to demands for higher control functionality, improved health monitoring and diagnostics, and better fuel efficiency. There are inherent limits to the capabilities of orthodox control systems to exploit the advantages of increasing microprocessor performance. As a result, there is now a pressing need to find alternative architectures to achieve the increase in computational performance necessary for future fault-tolerant control. New architectures will not only have to meet all existing requirements of safety, integrity, reliability and cost, but will also have to be capable of adapting to meet evolving needs. Modular, scaleable solutions based on parallel architectures offer many significant advantages over conventional approaches.

1.10.6 Robotics Applications

Parallel processing is an integral part of many robotics applications. Recent surveys show that autonomous systems will have an increasing market share in the future. The basic features needed by autonomous systems to support their activities are the ability for sensing, planning, and acting. These features enable a robot to act in its environment securely and to accomplish a given task. In dynamic environments, the necessary adaptation of the robot action is provided by closed control loops comprising sensing, planning, and acting. Unfortunately, this control loop could not be closed in dynamic environments, because of the long execution times of the single components. With this, the time intervals of single iterations become too large for a sound integration of the components into a single control loop. To pursue this approach, a reduction in the runtime is required, for which parallel computing is promising. In fact, industries are utilising parallel processing in various areas of robotics. Research is also being carried out in several establishments in the field of robotics using parallel computing.

1.10.7 Signal Processing and Control

One of the major areas of application of parallel processing is signal processing and control. Faster signal processing and speedy controls are prerequisites for many real-time applications, and these can be achieved through parallel processing. Indeed, there are continual developments in the fields of signal processing and control and as a result algorithms are becoming more and more complex with varying computational requirements. General-purpose processors or even purpose-built processors are not capable of meeting the real-time requirements of these algorithms. Accordingly, parallel processing is extensively utilised in these fields. Examples of such applications include filtering and analysis of radar and satellite signals in NASA, robotic arm control and industrial process control.

1.11 Algorithms and Architectures

It is an essential requirement in oder to gain maximum benefit from parallel processing to seek a close match between the algorithms to be parallelised and the parallel architectures to be used. Some algorithms are regular in nature, some are of irregular nature and some possess both regularities and irregularities. For example, a matrix-based algorithm would be regular in nature while most other signal processing algorithms generally possess irregularities. Complex algorithms used in the fields of signal processing and control might contain regularities and irregularities. It is required to form a parallel architecture in accordance with the nature of the application algorithm. The architecture could be either homogeneous or heterogeneous. A homogeneous architecture is one containing processors having the same characteristics and a heterogeneous architecture is one containing processors having different characteristics. For an algorithm of homogeneous nature a homogeneous architecture would be suitable to execute the algorithm while a heterogeneous architecture could be used to execute an algorithm of heterogeneous nature. When executing a heterogeneous algorithm using heterogeneous architecture, the regular parts of the algorithm can be allocated to processors that are designed to execute regular algorithms while the irregular parts can be assigned to processors that are designed to execute irregular algorithms.

1.12 Signal Processing and Control Algorithms

There are numerous signal processing and control algorithms in use in many fields, such as aerospace, radar, sonar, robotics, medical fields, automobiles, to name a few. The algorithms used in these fields range from fast Fourier transform (FFT) to Kalman filtering and proportional, integral, derivative (PID) controllers to neuro-fuzzy controllers. Many new algorithms are also being developed at a rapid pace. Complexity and shorter sample time are two of the main characteristics of such algorithms, which demand high-performance computing. General-purpose processors cannot satisfy these demands, although they are being developed at a rapid pace based on processing speed, communication ability and control ability. One of the alternative strategies in this case is to use specialised high-performance processing elements (PEs). For example, digital signal processing (DSP) devices are designed in hardware to perform concurrent add and multiply instructions and execute irregular algorithms efficiently. Vector processors are designed to process regular algorithms involving matrix manipulation efficiently. The ideal performance of these processors demands a perfect match between the capability of their architecture and the program behaviour. These include algorithm design, data structures, language efficiency, programmer's skill and compiler technology (Ching and Wu, 1989; Hwang, 1993; Tokhi *et al.*, 1997a,b, 1999a,b). However, in many cases purpose-built PEs are not efficient enough to bridge the gap between the hardware and the computing requirements of algorithms. This has resulted in resurgence in the development of parallel processing (PP) techniques to make the real-time realisation of modern signal processing and control algorithms feasible in practice.

1.13 Research Scope

A significant amount of research has been carried out in the field of parallel processing some of which is mentioned here. Special purpose processors such as DSPs and digital control processors have been reported to have been utilised by researchers in the early 1980s (Chen, 1982). However, real-time parallel implementations were adopted soon after and were boosted tremendously with the advent of transputers in 1987. Research in the area of real-time signal processing and control using transputer-based homogeneous architectures is widely reported in the literature (Fleming, 1988). Kourmoulis (1990) addressed the performance of homogeneous architectures comprising a number of transputers in the real-time simulation and control of flexible beam structures. Real-time implementation of Kalman filtering and self-tuning control algorithms utilising transputer-based homogeneous architectures were reported by Maguire (1991). Lawes and Clarke (1994) addressed the utilisation of 240 transputer nodes for real-time aircraft control. Tokhi and Hossain (1995) addressed real-time implementation of different signal processing and control algorithms using transputer-based parallel architectures. Ramos-Hernandez (1998) addressed parallel implementation of adaptive filtering algorithms using homogeneous parallel architectures incorporating a number of transputer nodes.

Homogeneous parallel architectures comprising DSPs were developed simultaneously with transputer-based homogeneous parallel architectures and are reported to have been used extensively in real-time signal processing and control applications (Costa, 1989; King and Picton, 1990). An investigation into the development of parallel adaptive/self-tuning controllers based on systollic arrays comprising a combination of transputers and DSPs was carried out by Li and Rogers (1991). Homogeneous parallel architectures consisting of a number of DSPs have also been reported to have been used by Tokhi *et al.* (1996) in implementing real-time beam simulation and least mean squares (LMS) adaptive filtering algorithms. Further research with homogeneous parallel architectures comprising DSPs was carried out by Madeira *et al.* (1998), where a Doppler signal spectral estimator was implemented on DSP-based parallel architectures.

A considerable amount of work has been carried out on heterogeneous parallel architectures for real-time signal processing and control. Baxter *et al.* (1994) proposed a generic method and a suit of design tools for the implementation of control algorithms on heterogeneous architectures. The tools are demonstrated by developing several case study algorithms to full implementation with an emphasis on the problematic areas leading to performance degradation common to parallel systems. Utilisation of heterogeneous parallel architectures in implementing real-time control algorithms giving special attention to partitioning and mapping complexity, shared memory problems and interprocessor communication, has been addressed by a number of researchers (Barragy *et al.*, 1994; Crummey *et al.*, 1994). High-performance real-time computing methods based on signal processing and control algorithms is addressed by Tokhi *et al.* (1997a,b). The optimisation problem for minimising the total execution time of an application program executed in a heterogeneous architecture with respect to matching, scheduling and data relocation is addressed by Tan and Siegel (1998).

As can be seen, research in the field of parallel processing covers a wide area. To speed up the process and meet the real-time requirements, algorithms used in various applications entail parallelisation. Thus, extensive research in this field continues. Many new processors and parallel architectures are being developed. There is a continued demand to investigate these processors and architectures to achieve maximum speedup and scaleup. Development of new performance metrics, task allocation strategies among the processors, reduction of communication overhead, synchronising the tasks are several other important research areas worth mentioning.

1.14 Summary

Parallel processing is becoming increasingly important in various applications. It is utilised in many fields for better speedup and scaleup. To gain maximum benefit from a parallel solution requires various issues to be considered. It is always necessary to do some kind of background study and investigation on the basis of the factors involved to find whether parallel processing is worthwhile for a particular case. Currently, parallel processing is used extensively in various applications. Among all such applications signal processing and control demand special attention due to their general applicability in various fields and the current demand for automation.

1.15 Exercises

1. Define and describe parallel processing with real-life examples not mentioned in this chapter? Why is parallel processing necessary in various applications?

2. What is meant by real-time systems? What are the forms of real-time systems? Describe why parallel processing is necessary for some real-time problems.

3. Describe the components and characteristics of parallel processing.

4. What are the various levels of parallel processing? Describe them briefly. Describe the various steps involved in realising parallel processing.

5. What is meant by speedup and scaleup in relation to parallel processing? Describe a few major benefits of parallel processing.

6. What are the various factors to be considered when implementing a parallel processing scheme? Describe a few of these factors.

7. What is meant by communication overhead? How does it affect the performance of parallel processing?

8. How does the amount of inherent parallelism affect the performance of parallel processing?

9. What is meant by message passing? How does it take place?

10. Briefly explain some of the major applications of parallel processing.

11. Why is parallel processing so important in signal processing and control applications?

12. Explain the necessity for matching algorithms and architectures from the perspective of parallel processing.

2. Parallel Architectures

2.1 Objectives

- To introduce the principles and classification of parallel architectures.
- To discuss various forms of parallel processing.
- To explore the characteristics of parallel architectures.

2.2 Introduction

Parallel processing has emerged as an area with the potential of providing satisfactory ways of meeting real-time computation requirements in various applications and the quest for speed, which is a natural tendency of human beings, as discussed in Chapter 1. As such, various forms of parallel architectures have evolved and research is being carried out worldwide to discover new architectures that can perform better than existing architectures. With speed as the prime goal, the factors generally considered in developing a new architecture include the internal circuitry of processors or PEs, number of PEs, arrangement of PEs and memory modules in an architecture, the communication mechanism among the PEs and between the PEs and memory modules, number of instruction and data streams, nature of memory connections with the PEs, nature and types of interconnection among the PEs, and program overlapping. The internal circuitry of PEs plays a vital role in designing parallel architecture. Some parallel architectures are designed with small number of PEs of complex internal circuitry to enhance the overall performance of the architecture. Other architectures, on the other hand, are designed with a substantial number of PEs of simple internal circuitry to achieve the desired performance. The performance of an architecture can often be increased by adding more PEs. However, this is possible only up to a certain limit, as adding more PEs to the architecture incurs more communication overhead and it may not be cost effective beyond a limit. The arrangement of processors is also crucial to the design of a parallel architecture. Processors can be arranged in different forms using various types of interconnection strategies, such as static and dynamic. In a static form, the format of the architecture remains fixed and is not expandable by adding more processors, while in a dynamic format more processors can be added under system control to meet a particular requirement. The communication mechanism among PEs in some architectures is straightforward, while in others the communication scheme is complicated and requires extra effort and circuitry. Similar to the arrangement of PEs, the arrangement of memory modules in an architecture is important and contributes to the development of varied forms of parallel architectures. The memory arrangement in some architectures is global, i.e., all PEs use a common memory or memory modules, which helps to establish communication among the processors, while others have memory modules associated with individual PEs and communication is established via messages passing among the PEs. Some forms of parallel architecture have evolved on the basis of the methodology an algorithm is implemented. A pipeline architecture is an example of this. There are also special forms of parallel architectures developed on the basis of merging the characteristics of various forms of existing parallel architectures. Many attempts have also been made to develop application-specific parallel architectures. For example, vector parallel architectures have been developed to execute vector intensive algorithms, DSP devices have been designed for efficient implementation of signal processing, e.g. digital filtering algorithms.

Indeed, research and development in parallel computing is a continuing effort, and it is accordingly expected that various forms of parallel architectures will emerge in the future. Whatever may be the basis of development, each type of parallel processor has its own characteristics, advantages, disadvantages and suitability in certain application areas.

2.3 Classifications

A vast number parallel architecture types have been devised over the years. Accordingly, it is not easy to develop a simple classification system for parallel architectures. Moreover, various types of parallel architecture have overlapping characteristics to different extents. However, the various forms of parallel architecture can be distinguished under the following broad categories:

- Flynn's classification.
- Classification based on memory arrangement and communication among PEs.
- Classification based on interconnections among PEs and memory modules.
- Classification based on characteristic nature of PEs.
- Specific types of parallel architectures.

2.3.1   Flynn's Classification

Michael J. Flynn introduced a broad classification of parallel processors based on the simultaneous instruction and data streams during program execution (Flynn, 1966). It is noted here that instruction and data streams are two main steps that occur during program execution, as depicted in Figure 2.1. It shows that during program execution the PE fetches instructions and data from the main memory, processes the data as per the instructions and sends the results to the main memory after processing has been completed. The instructions are referred to as an instruction stream, which flows from the main memory to the PE, and the data is referred to as the data stream, flowing to and from the PE. Based on these streams, Flynn categorised computers into four major classes, which are described below.

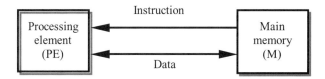

Figure 2.1. Instruction stream and data stream

Single-instruction Single-data Stream
The simple Von Neumann computer shown in Figure 2.2 falls under the category of single-instruction single-data (SISD) stream. An alternative representation of the architecture indicating instruction and data flow is shown in Figure 2.3. The SISD computer architecture possesses a central processing unit (CPU), memory unit and input/output (I/O) devices. The CPU consists of an arithmetic and logic unit (ALU) to perform arithmetic and logical operations, control unit (CU) to perform control operations and registers to store small amounts of data. SISD computers are sequential computers and are incapable of performing parallel operations.

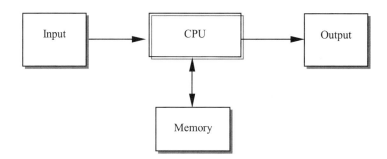

Figure 2.2. Von Neumann (SISD) computer architecture

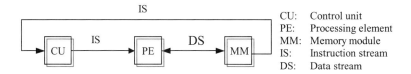

Figure 2.3. SISD architecture with instruction and data flow

Single-instruction Multiple-data Stream

The general structure of a single-instruction multiple-data (SIMD) parallel architecture is shown in Figure 2.4. This architecture possesses a single instruction stream, to process the entire data structure (multiple-data stream). In other words, in this configuration, a single program control unit or control processor controls multiple execution units or execution processors. Each execution processor is equipped with a local memory to store the data it will work on. Each execution processor executes the same instruction issued by the control processor on its local data. The execution processors are capable of communicating with each other when required. SIMD architectures are good for problems where the same operation is executed on a number of different objects, for example image processing. Some other suitable applications of SIMD architectures include matrix operations and sorting. Programming is quite simple and straightforward for this architecture. SIMD architecture could be divided into two subclasses according to the interconnections between the PEs. These are: vector architecture and array architecture.

Organisation of the SIMD vector architecture is shown in Figure 2.5. The PEs are connected to each other via special data links. These links are used to perform simple data exchange operations like shifts and rotations (Hays, 1988). All the PEs are connected to the central control processor to obtain the instructions to be executed.

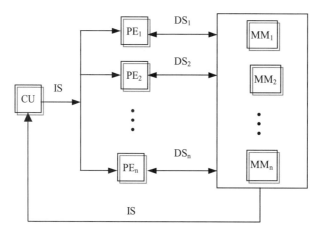

Figure 2.4. General structure of SIMD computer architecture

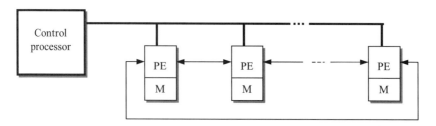

Figure 2.5. Vector architecture

The SIMD computer is often used as a synonym for array architecture. Unlike vector architecture, PEs in an array architecture are connected by interconnecting networks. A general form of array structure is shown in Figure 2.6, where a two-dimensional grid of PEs executes instructions provided by the control processor. Each PE is connected to its four neighbours to exchange data. End-around connections also exist on both rows and columns, which are not shown in Figure 2.6. Each PE is capable of exchanging values with each of its neighbours. Each PE possesses a few registers and some local memory to store data. Each PE is also equipped with a special register called a network register to facilitate movement of values to and from its neighbours. Each PE also contains an ALU to execute arithmetic instructions broadcast by the control processor. The central processor is capable of broadcasting an instruction to shift the values in the network registers one step up, down, left or right (Hamacher *et al.*, 2002). Array architecture is very powerful and will suit problems that can be expressed in matrix or vector format.

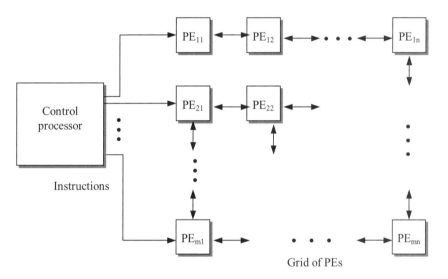

Figure 2.6. Array architecture

Both array and vector architectures are specialised machines. They are mainly designed to solve numerical problems comprising substantial numbers of vector and/or matrix elements. The basic difference between vector and array architectures is that high-performance is achieved in vector architecture through exploiting a pipelining mechanism whereas in array architecture a large number of PEs are incorporated to boost the performance. None of the architectures is well suited to enhance the performance of general computation. The most famous example of an array processor is ILLIAC-IV designed at the University of Illinois and built by Burroughs Corporation. It was developed using 64 processors. Other important examples of array processors include the Thinking Machine Corporation's CM-2 processor, which could have up to 65536 processors, Maspar's MP-1216 processors, which could accommodate 16384 processors, and the Cambridge parallel processing Gamma II plus machine, which could accommodate up to 4096 processors (Hamacher *et al.*, 2002).

Multiple-instruction Single-data Stream
The general structure of a multi-instruction single-data stream (MISD) architecture is shown in Figure 2.7. This architecture possesses a multiple instruction stream and single data stream. This architecture has not evolved substantially and thus, there are not many examples of the architecture (Hays, 1988). However, a major class of parallel architectures, called pipeline computers can be viewed as MISD architecture. Pipeline computers will be discussed later in this chapter.

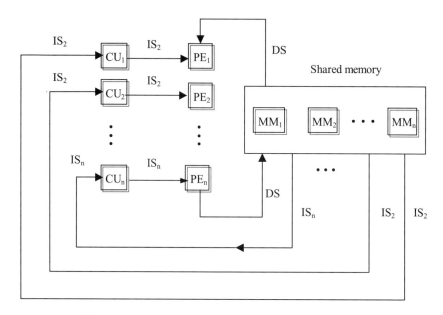

Figure 2.7. General structure of MISD computer architecture

For most applications, MISD computers are rather awkward to use, but can be useful in applications of a specialised nature (Akl, 1989; Kourmoulis, 1990). A typical example of one such specialised application is robot vision. For example, a robot inspecting a conveyor belt by sight may have to recognise objects that belong to different classes. An MISD machine can quickly carry out a classification task by assigning each of its processors a different class of objects and after receiving what the robot sees each processor may carry out tests to determine whether the given object belongs to its class or not.

Multiple-instruction Multiple-data Stream
Figure 2.8 shows the general structure of multiple-instruction multiple-data (MIMD) stream architecture. This architecture is the most common and widely used form of parallel architectures. It comprises several PEs, each of which is capable of executing independent instruction streams on independent data streams.

The PEs in the system typically share resources such as communication facilities, I/O devices, program libraries and databases. All the PEs are controlled by a common operating system. The multiple PEs in the system improves performance and increase reliability. Performance increases due to the fact that computational load is shared by the PEs in the system. Theoretically, if there are n PEs in the system, the performance will increase by n times in comparison to a single PE based system. System reliability is increased by the fact that failure of one PE in the system does not cause failure of the whole system.

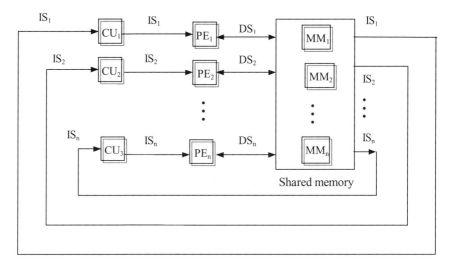

Figure 2.8. General structure of MIMD computer architecture

2.3.2 Classification Based on Memory Arrangement and Communication among PEs

Parallel architectures can be classified into two major categories in terms of memory arrangement. These are: shared memory and message passing or distributed memory. In fact, these architectures constitute a subdivision of MIMD parallel architecture. Shared memory and distributed memory architectures are also called tightly coupled and loosely coupled architectures respectively. Each type of architecture has its advantages and disadvantages.

Shared Memory Multiprocessor
In a shared memory multiprocessor configuration multiple processors share a common memory unit comprising a single or several memory modules. All the processors have equal access to these memory modules and these memory modules are seen as a single address space by all the processors. The memory modules store data as well as serve to establish communication among the processors via some bus arrangement. Communication is established through memory access

instructions. That is, processors exchange messages between one another by one processor writing data into the shared memory and another reading that data from the memory.

Programming this architecture is quite straightforward and attractive. The executable programming codes are stored in the memory for each processor to execute. The data related to each program is also stored in this memory. Each program can gain access to all data sets present in the memory if necessary. The executable codes and shared data for the processor can be created and managed in different ways by designing parallel programming language or using existing sequential languages such as C/C++. There is no direct processor-to-processor communication involved in the programming process; instead communication is handled mainly via the shared memory modules. Access to these memory modules can easily be controlled through an appropriate programming mechanism such as multitasking. However, this architecture suffers from a bottleneck problem when a number of processors endeavour to access the global memory at the same time. This limits the scalability of the system. As a remedy to this problem most large, practical shared memory systems have some form of hierarchical or distributed memory structure such that processors can access physically nearby memory locations faster than distant memory locations. This is called non-uniform memory access. Figure 2.9 shows a general form of shared memory multiprocessor architecture.

Shared memory multiprocessors can be of two types, namely uniform memory access (UMA) architecture and non-uniform memory access (NUMA) architecture.

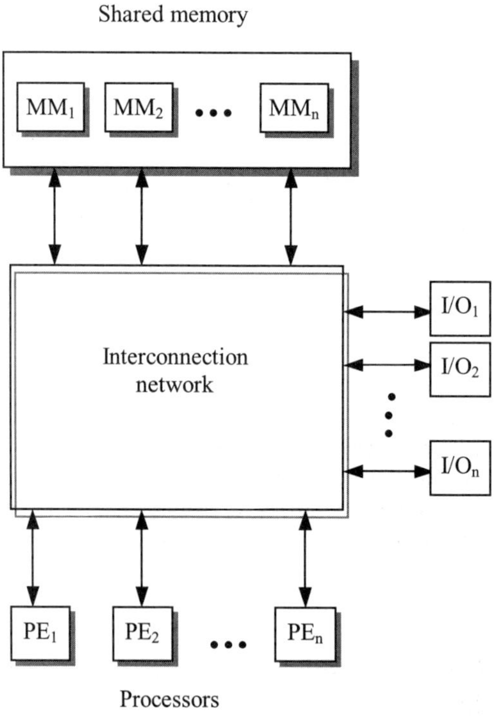

Figure 2.9. Shared-memory multiprocessor

As the name suggests, the memory access time to the different parts of the memory are almost the same in the case of UMA architectures. UMA architectures are also called symmetric multiprocessors. An UMA architecture comprises two or more processors with identical characteristics. The processors share the same main memory and I/O facilities and are interconnected by some form of bus-based interconnection scheme such that the memory access time is approximately the same for all processors. All processors can perform the same functions under control of an integrated operating system, which provides interaction between processors and their programs at the job, task, file and data element levels (Stallings, 2003). The IBM S/390 is an example of UMA architecture.

In the case of NUMA architectures the memory access time of processors differs depending on which region of the main memory is accessed. A subclass of NUMA system is cache coherent NUMA (CC-NUMA) where cache coherence is maintained among the caches of various processors. The main advantage of a CC-NUMA system is that it can deliver effective performance at higher levels of parallelism than UMA architecture.

Message Passing Multicomputer
A distributed memory architecture is different from a shared memory architecture in that each unit of this architecture is a complete computer building block including the processor, memory and I/O system. A processor can access the memory, which is directly attached to it. Communication among the processors is established in the form of I/O operations through message signals and bus networks. For example, if a processor needs data from another processor it sends a signal to that processor through an interconnected bus network demanding the required data. The remote processor then responds accordingly. Certainly, access to local memory is faster than access to remote processors. Most importantly, the further the physical distance to the remote processor, the longer it will take to access the remote data. On one hand this architecture suffers from the drawback of requiring direct communication from processor to processor, on the other hand, the bottleneck problem of shared memory architecture does not exist. A general form of shared memory architecture is shown in Figure 2.10.

The speed performance of distributed memory architecture largely depends upon how the processors are connected to each other. It is impractical to connect each processor to the remaining processors in the system through independent cables. It can work for a very low number of processors but becomes nearly impossible as the number of processors in the system increases. However, attempts have been made to overcome this problem and as a result several solutions have emerged. The most common of these is to use specialised bus networks to connect all the processors in the system in order that each processor can communicate with any other processor attached to the system.

2.3.3 Classification Based on Interconnections between PEs and Memory Modules

Parallel architectures are also classified in terms of interconnecting network arrangements for communication among the various PEs included in the architecture. In fact, this classification is quite specific to MIMD architectures as they, generally, comprises multiple PEs and memory modules. The various interconnecting communication networks used for establishing communication schemes among the PEs of a parallel architecture include: linear, shared single bus, shared multiple bus, crossbar, ring, mesh, star, tree, hypercube and complete graph. Among these interconnecting networks, linear, mesh, ring, star, tree, hypercube and complete graph are static connection structures whereas shared single bus, shared multiple bus and crossbar are dynamic interconnection structures as they are reconfigurable under system control (Hays, 1988).

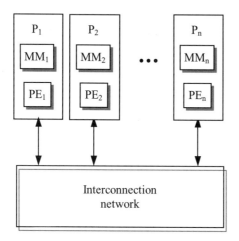

Figure 2.10. Distributed-memory multiprocessor

Linear Network
A number of nodes are connected through buses in a linear format to form a network of nodes as shown in Figure 2.11. Every node, except the nodes at the two ends, in this configuration is directly connected to two other nodes. Thus, to connect n nodes in this configuration $n-1$ buses are required and the maximum internodes distance is $n-1$.

Figure 2.11. Linear interconnection structure

Single Shared Bus Network
The single shared bus interconnection structure, shown in Figure 1.12, is widely used in parallel architectures. A number of PEs and memory units are connected to a single bus in this case, through which communication is established among the PEs and memory units connected to it.

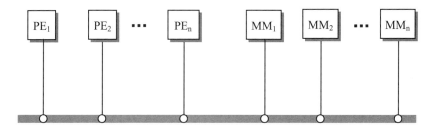

Figure 2.12. Single bus interconnection structure

The operation of the single bus interconnection structure, in its simplest form, is as follows: when a processor issues a read request to a memory location it holds the bus until it receives the expected data from the memory module. It will require some time for the memory module to access the data from the appropriate location. The transfer scheme also will need some time and another request from any processor will not be initiated until the transfer is completed, which means the bus will remain idle for a certain amount of time that can be as high as two-thirds of the total time required for the transfer. This problem has been overcome by using a split transfer protocol scheme whereby the bus can handle a number of requests from different processors to different memory modules. In this case after transferring the address of the first request the bus starts the transfer of the next request so that two requests are executed in parallel. If none of the two requests has completed, the bus can be assigned a third request. When the first memory module completes its access cycle, the bus is used to send the data to the destination processor. As another module completes its access cycle the data is transferred using the bus, and the process continues. The split transfer protocol increases the performance of the bus at the cost of complexity. The complexity increases due to maintaining synchronisation and coordination among the requests, processors and memory modules.

One of the limitations of single bus interconnection is that a large number of processors and memory modules cannot be connected to a bus. The number of modules to be connected with the bus could be increased by using a wider bus with increased bandwidth. However, the bandwidth of a single bus is limited by the connection for the use of the bus and by the increased propagation delays caused by the electrical loadings when many modules are connected (Hamacher *et al.*, 2002).

Multiple Shared Buses Network
The single shared bus network cannot cope with large numbers of PEs and memory units as mentioned above. Multiple shared bus networks are used in such cases. The structure of a multiple shared bus network is shown in Figure 2.13, where each processor and memory are connected to one or more of the available buses, each of which possesses all the attributes of an independent system bus. Besides reducing the communication load per bus, a degree of fault tolerance is provided, since the system can be designed to continue operation, possibly with reduced performance, if an individual bus fails (Hays, 1988).

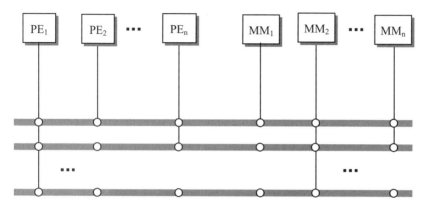

Figure 2.13. Multiple buses interconnection structure

Crossbar Interconnection Network
The structure of crossbar architecture is shown in Figure 2.14. In this architecture, all the PEs and memory modules are interconnected through a multibus crossbar network system where subscript m denotes the memory and n denotes the PEs. The crossbar architecture becomes very complex as the number of memory modules and PEs increases.

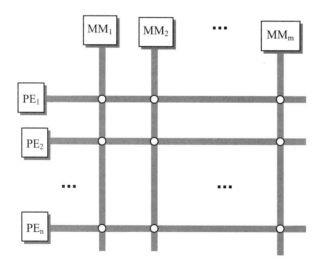

Figure 2.14. Crossbar interconnection structure

Star Interconnection Network
The star interconnection, as shown in Figure 2.15, is one of the simplest interconnection networks. In this configuration $n-1$ buses are required to connect

n nodes and the maximum internode distance is 2. A node in this structure can communicate with any other node through the node in the centre.

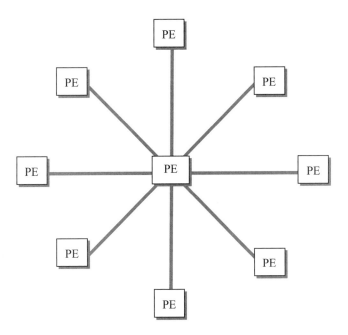

Figure 2.15. Star interconnection structure

Ring Interconnection Network
The ring network, shown in Figure 2.16, is also one of the simplest interconnection topologies. This interconnection is very easy to implement. In the case of ring interconnection *n* buses are required to connect *n* nodes and the maximum internodes distance is $n/2$. Rings can be used as building blocks to form other interconnection structures such as mesh, hypercube and tree. A ring-based two-stage tree structure is shown in Figure 2.17. However, the highest-level ring could be a bottleneck for traffic in this case. Commercial machines such as Hewlett-Packard's Exemplar V2600 and Kendal Square Research's KSR-2 have been designed using ring networks.

Tree Interconnection Network
Tree structure is another important and useful interconnection topology. There could be a number of levels in a tree structure. The general form of an *n*-level tree structure is shown in Figure 2.18. In this case any intermediate node acts as a medium to establish communication between its parents and children. Through this mechanism communication could also be established between any two nodes in the structure. A tree structure can be highly effective if a small portion of traffic goes through the root node otherwise due to bottleneck problems performance

deteriorates rapidly. The possibility of bottleneck problems is less in a flat tree structure where there is a large number of nodes at the higher levels.

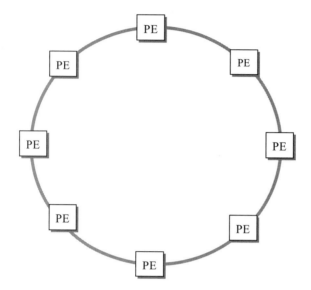

Figure 2.16. Ring interconnection structure

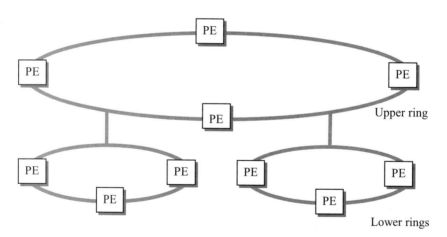

Figure 2.17. Two-stage tree networks based on ring networks

Hypercube Interconnection Network

Hypercube is a popular interconnection network architecture, especially for NUMA multiprocessors. An n-dimensional hypercube can connect 2^n nodes each of which includes a processor, a memory module and some I/O capability. A three-dimensional hypercube is shown in Figure 2.19. The edges of the cube represent bi-directional communication links between two neighbouring nodes. The nodes are normally labelled using binary addresses in a way that the addresses of the two neighbouring nodes differ by one bit position. Transferring messages from one node to another in a hypercube structure is accomplished with the help of binary addresses assigned to each of the nodes. In this transferring scheme the binary address of the source node and the destination nodes are compared from least to most significant bits and transfer to the destination is performed through some intermediate nodes in between. For example, the transfer of message from node P_i to a node P_j takes place as follows. First the binary addressees of P_i and P_j are compared from least to most significant bits. Suppose they differ in bit position p. Node P_i then sends a message to the neighbouring node P_k whose address differs from P_i in bit position p. Node P_k then passes the message to its appropriate neighbours using the same scheme. The message gets closer to the destination node with each of these passes and finally reaches it after several passes. Consider, for example, that node P_3 in Figure 2.19 wants to send a message to node P_6. It will require two passes through node P_2.

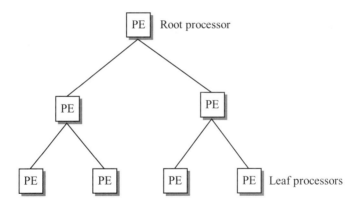

Figure 2.18. Tree interconnection structure

The hypercube structure is very reliable. If a faulty link is detected while passing a message from source to destination node through the shortest route; the message can be passed using another route. A hypercube is homogeneous in nature, as the system appears the same when viewed from any of its outside nodes. Thus, programming the hypercube is simple because all nodes can execute the same programs on different data when collaborating on a common task (Hays, 2003).

Many commercial multiprocessors have used hypercube interconnections including the Intel iPSC. A seven-dimensional hypercube has been used in this machine using 128 nodes. The NCUBE's NCUBE/ten used 1024 nodes in a 10-dimensional hypercube. However, the hypercube structure has lost much of its popularity since the advent of the mesh interconnection structure as an effective alternative (Hamacher *et al.*, 2002).

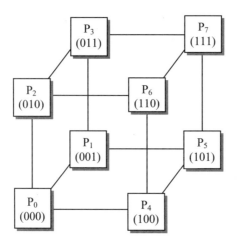

Figure 2.19. Hypercube interconnection structure

Mesh and Torus Interconnection Network
Mesh is a popular interconnection network structure used to connect large numbers of nodes. It came into being in the 1990s as an alternative to hypercube in large multiprocessors. A 16-node mesh structure is shown in Figure 2.20. To formulate a mesh structure comprising n nodes $2(n - n^{0.5})$ buses are required and the maximum internodes distance is $2(n^{0.5} - 1)$. Routing in a mesh is established in various ways. One of the simplest and most popular ways is to choose the path between a source node n_i and a destination node n_j then proceed in the horizontal direction from n_i to n_j. When the column in which n_j resides is reached the transfer proceeds in the vertical direction along that column. The Intel's Paragon is a well-known mesh-based multiprocessor. If a wraparound connection is made between the nodes at opposite edges the result is a network that consists of a set of bi-directional rings in the X direction connected by a similar set of rings in the Y direction. This network is called a Torus (Hamacher *et al.*, 2002). The average latency in a torus is less than in a mesh at the expense of complexity. Fujitsu's AP3000 is a torus connection based machine.

Complete Graph Interconnection Network
In a complete graph interconnection structure several processors are connected in the complete graph format as depicted in Figure 2.21. Here, each node can directly communicate with any other node without going through or touching any intermediate node. However, it requires many buses. For a complete graph with n nodes the number of buses required is $n(n-1)/2$ and the maximum internode distance is 1.

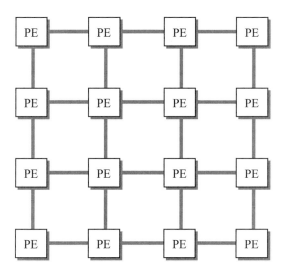

Figure 2.20. Mesh interconnection structure

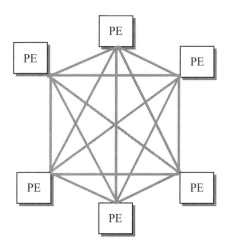

Figure 2.21. Complete graph interconnection structure

Switching or Dynamic Interconnection Structures
Dynamic parallel architectures are reconfigurable under system control and the control is generally achieved through different kinds of switching circuits. One such switch is shown in Figure 2.22, which is an AND gate controlling the connection between two lines namely m and n. When line n is high (say, a binary 1) indicating that a connection is required to be made with the line m, the control line will go high and the connection will be established.

Another two-state switching element is shown in Figure 2.23. Each switch has a pair of input buses x_1 and x_2, and a pair of output buses y_1 and y_2, assisted by some form of control mechanism. The buses connected to the switch could be used to establish processor-to-processor or processor-to-memory links. The switch S has two states, determined by a control line, the through or direct state, as depicted in Figure 2.23, where $y_1 = x_1$ (i.e., y_1 is connected to x_1) and $y_2 = x_2$ and a cross state where $y_1 = x_2$ (i.e., y_1 is connected to x_2) and $y_2 = x_1$. Using S as a building block, multistage switching networks of the type can be constructed for use as interconnection networks in parallel computers.

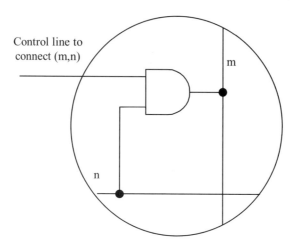

Figure 2.22. A typical crossbar switch

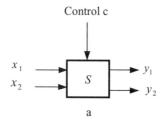

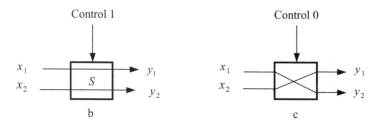

Figure 2.23. Two-states switching mechanisms

A three-stage switching network of this type is shown in Figure 2.24. The network contains 12 switching elements and is intended to provide dynamic connections between the processors. By setting the control signals of the switching elements in various ways, a large number of different interconnection patterns is possible (Hays, 1988). The number of stages, the fixed connections linking the stages, and the dynamic states of the switching elements, in general, determines the possibilities.

A comparison of features of a selected set of interconnection structures is given in Table 2.1.

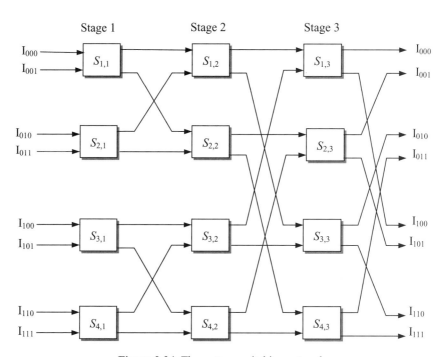

Figure 2.24. Three-stage switching network

2.3.4 Classification Based on Characteristic Nature of Processing Elements

Parallel architectures are also classified in terms of the nature of the PEs comprising them. An architecture may consist of either only one type of PE or various types of PEs. The different types of processors that are commonly used to form parallel architectures are described below.

CISC Processors
The acronym CISC stands for Complex Instruction Set Computer. It is a type of processor that uses a complex, powerful instruction set capable of performing

many tasks including memory access, arithmetic calculations and address calculations. Some distinctive features of CISC processors are as follows:

Table 2.1. Comparison of features of selected interconnection structures

Network type	Connections/PE	Maximum distance	Bandwidth	Scalability
Bus	1	2	Low	Poor
Crossbar	2	2	High	Good
Ring	2	$n/2$	Low	Good
Complete graph	$n-1$	1	High	Poor
Torus	4	$\sqrt{n}$	Good locally	Good
Hypercube	$\log_2 n$	$\log_2 n$	Good	Good

- CISC instruction sets are large and powerful.
- CISC instructions are executed slowly as each instruction is normally capable of doing many things.
- CISC processors are comparatively difficult to program.
- CISC architectures have pipelines and more registers.
- CISC processors handle only a relatively low number of operations.

CISC processors are generally used in desktop machines. The Motorola 68x0 and the Intel 80x86 families are examples of CISC processors.

RISC Processors
The abbreviation RISC stands for Reduced Instruction Set Computer. RISC processors have a number of distinguishing characteristics, some of which are as follows:

- RISC processors handle more operations than CISC processors.
- Execution of instructions in a RISC processor is faster than in their CISC counterpart.
- RISC processors support pipelined instruction execution.
- RISC processors contain large number of registers, most of which can be used as general-purpose registers.
- RISC processors are simple to program.

Current RISC processors include the M600-series PowerPC (Motorola/IBM), i960 (Intel), SPARC (Sun), ARM (Advanced RISC Machines), and Am 29000-series (Advanced Micro Devices).

DSP and Vector Processors

DSP chips are specially designed to execute DSP algorithms and applications such as FFT, correlation, convolution and digital filtering. Such algorithms are used extensively in a variety of DSP applications such as radar, sonar, and weather forecasting. As most DSP operations require additions and multiplications together, DSP processors usually possess adders and multipliers, which can be used in parallel within a single instruction. DSP chips are also capable of handling multiple memory access in a single instruction cycle. One of the major differences between DSP chips and general-purpose processors is that DSP chips are required to deal with real-world problems frequently and they are designed to do so. TMS320C4x, DSP563xx, and DSP96002 are examples of DSP chips.

Vector processors are designed to execute vector-intensive algorithms faster than other types of general-purpose and specialised processors. In fact, many algorithms are of regular nature and contain numerous matrix operations. Vector processors are very efficient at executing these types of algorithms. Examples of vector processors are the Intel i860 and i960.

Homogeneous and Heterogeneous Parallel Architectures

In a conventional parallel system all the PEs are identical. This architecture can be regarded as homogeneous. Figure 2.25 shows the homogeneous architecture of DSP chips and Figure 2.26 shows the homogeneous architecture of vector processors. However, many algorithms are heterogeneous, as they comprise functions and segments of varying computational requirements. Thus, heterogeneous architectures are designed to incorporate diverse hardware and software components in a heterogeneous suite of machines connected by a high-speed network to meet the varied computational requirements of a specific application (Tan and Siegel, 1998). In fact, heterogeneous architectures represent a more general class of parallel processing system. The implementation of an algorithm on a heterogeneous architecture, having PEs of different types and features, can provide a closer match with the varying computing requirements and, thus, lead to performance enhancement. A typical heterogeneous architecture is shown in Figure 2.27, which comprises RISC processors, DSP processors and a vector processor.

2.3.5 Specific Types of Parallel Architectures

Various forms of parallel processors are evolving to cope with complex algorithms and short sample time requirements. Some of the specialised forms of parallel processors are described below.

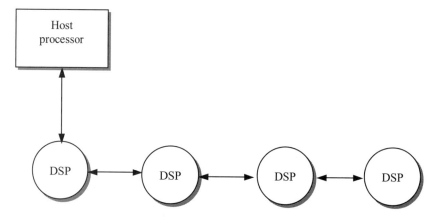

Figure 2.25. Homogeneous architecture of DSP processors

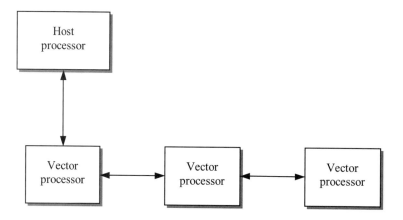

Figure 2.26. Homogeneous architecture of vector processors

Pipeline Architecture

Pipeline is a very widely used parallel architecture, designed to execute pipelined instructions. Pipeline is an MISD type processor. However, it could be of MIMD type as well, depending upon the structures and operations.

In the case of pipeline execution while one instruction is executed, the next instruction in the sequence is decoded, while a further one is fetched. The processor consists of a sequence of stages and the operands are partially executed at each stage and the fully processed result is obtained after the operands have passed through all the stages. A three-stage pipelined processing mechanism is shown in Figure 2.28. As shown, when operand 3 is being executed after having been fetched and decoded, operand 2 is being decoded after having been fetched and operand 1 is being fetched. All stages are busy at all times. In contrast in sequential processing when one stage is busy the other two remain idle.

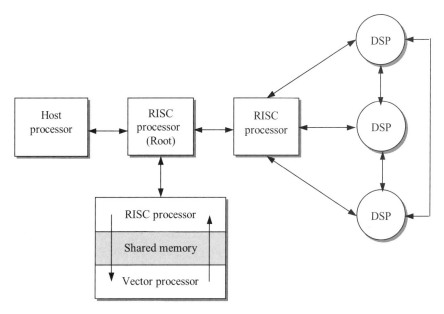

Figure 2.27. Heterogeneous architecture

Execute				Operand 3

Figure 2.28. Three-stage pipeline processing

A pipeline processor consists of a number of stages called segments, each segment comprising an input register and a processing unit. An n-stage pipeline processor is shown in Figure 2.29. The registers play their role as buffers compensating for any differences in the propagation delays through the processing units (Hays, 1988). Generally, the whole process is controlled by a common clock signal. All the registers change their state synchronously at the start of a clock period of the pipeline. Each register then receives a new set of data from the preceding segment except the first register, which receives data from an external source. In each clock period, all the segments transfer their processed data to the next segment and compute a new set of results.

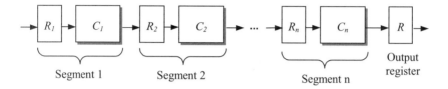

Figure 2.29. A general n-stage pipeline processing structure

From the operand point of view, pipeline is categorised into two types, namely, the instruction pipeline and arithmetic pipeline. Instruction pipelines are designed to speed up the program control functions of a computer by overlapping the processing of several different instructions namely fetch, decode and execute. Arithmetic pipelines are designed to execute special classes of operands very fast. These arithmetic operations include, multiplication, floating-point operations and vector operations.

Example of an Arithmetic Pipeline
The concepts of instruction and arithmetic pipelining are similar. However, at the implementation level an arithmetic pipeline is relatively complex. To develop an understanding of pipelining, an example arithmetic pipeline for floating-point addition is illustrated here. Figure 2.30 shows a five-segment floating-point adder pipeline, where a denotes a sequence of floating-point numbers, a_M denotes the mantissa of the sequence and a^E denotes the exponent of the sequence. b denotes another sequence of floating-point numbers with b_M and b^E the mantissa and exponent, respectively. Let two sequences of floating point (normalised) numbers be added using a five-segment floating-point adder, as shown in Figure 2.30. An example of a five-segment floating-point operation is shown in Figure 2.31, where each of the five segments can contain a pair of partially processed scalar operands (a_i, b_i). Buffering in the segments ensures that S_i only receives, as inputs, the results computed by segments S_{i-1} during the preceding clock period. If the pipeline clock period is T seconds long, i.e., the execution time of each segment, then it takes a total time of XT to compute a single sum $a_i + b_i$, where $X(= 5)$ represents the number of segments. This is approximately the time required to do one floating-point addition using a non-pipelined processor, plus the delay due to the buffer registers. Once all five segments of the pipeline are filled with data, a new sum emerges from the fifth segment every T seconds. Figure 2.32 shows the time space diagram for the process for the first 10 clock cycles. Thus, the time required to perform N consecutive additions can be calculated as follows:

It follows from the time space diagram in Figure 2.32 that the time required to fill all five segments is $(5-1)T = 4T$, therefore for X segments this will be $(X-1)T$, and the total execution time required to compute N operations will be $NT + (X-1)T$, implying that the pipeline's speedup is:

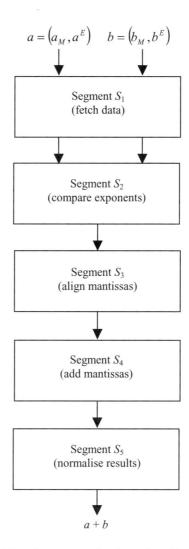

$$a = \left(a_M, a^E\right) \quad b = \left(b_M, b^E\right)$$

Segment S_1
(fetch data)

Segment S_2
(compare exponents)

Segment S_3
(align mantissas)

Segment S_4
(add mantissas)

Segment S_5
(normalise results)

$a + b$

Figure 2.30. Five-segment floating-point adder pipeline

$$S(X) = \frac{NXT}{NT + (X-1)T} = \frac{NX}{N + X - 1}$$

For large N the above approximates to $S(X) \approx X$. It is therefore clear that a pipeline with X segments is X-times faster than a non-pipelined adder.

Figure 2.33 shows the equivalent block representation of the five-segment floating-point adder pipeline in Figure 2.30, using combinational circuits. Suppose, the time delays of the four segments are $t_1 = 60\,\text{ns}$, $t_2 = 70\,\text{ns}$, $t_3 = 100\,\text{ns}$,

$t_4 = 80\,\text{ns}$, $t_5 = 80\,\text{ns}$ and the interface registers have a delay of $t_r = 10\,\text{ns}$. The clock period is chosen as $t_p = t_3 + t_r = 110\,\text{ns}$. An equivalent nonpipeline floating-point adder will have a delay time $t_n = t_1 + t_2 + t_3 + t_5 + t_r = 400\,\text{ns}$. In this case, the pipelined adder has a speedup of $400/110 = 3.64$ over the non-pipelined adder.

Consider two floating-point numbers,

$a_1 = 0.9504 \times 10^3$, where, mantissa is *0.9504* and exponent is *3*

and , $b_1 = 0.8200 \times 10^2$, where, mantissa is *0.8200* and exponent is *2*. The activities of the different segments will be as follows:

Segment 1: Fetch the values of *a* and *b*.
Segment 2: Compare the exponent of the two numbers. Consider, the larger exponent
(which is 3) as the exponent of the result.
Segment 3: Align the mantissa of b_1 for exponent 3 giving

$a_1 = 0.9504 \times 10^3$

$b_1 = 0.08200 \times 10^3$.

Segment 4: Add mantissas of the two numbers, giving
0.9504 + 0.0820 = 1.0324, which is the mantissa of the result.
Thus, the result before normalisation will be $c_1 = 1.0324 \times 10^3$.
Segment 5: Finally, normalise the result giving
$c_1 = 0.10324 \times 10^4$, available at the output of the pipeline.

Figure 2.31. An example for operations of a five-segments floating-point adder

Clock cycle	1	2	3	4	5	6	7	8	9	10
Segment 1	T_1	T_2	T_3	T_4	T_5	T_6				
Segment 2		T_1	T_2	T_3	T_4	T_5	T_6			
Segment 3			T_1	T_2	T_3	T_4	T_5	T_6		
Segment 4				T_1	T_2	T_3	T_4	T_5	T_6	
Segment 5					T_1	T_2	T_3	T_4	T_5	T_6

Figure 2.32. Time space diagram

Exponents   Mantissas

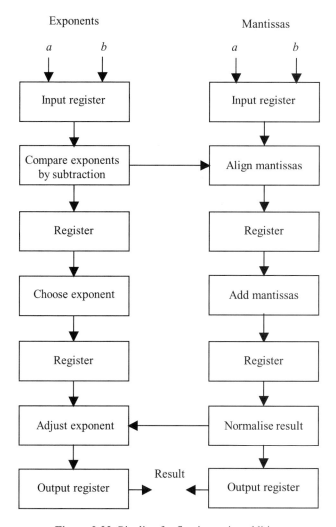

Figure 2.33. Pipeline for floating-point addition

Problem 2.1: *Consider a four-segment pipeline processor executing 200 tasks. Let the time it takes to process a sub-operation in each segment be 20 ns . Determine the speedup of the pipeline system.*

Solution: The execution time for a non-pipeline system will be

$$4NT = 4 \times 200 \times 20 = 16000 \, \text{ns}$$

where number of tasks $N = 200$ and sub-operation process time for each segment $T = 20 \, \text{ns}$.

Execution time for a pipeline system will be,

$$(N + 3)T = (200 + 3) \times 20 = 4060 \, \text{ns}$$

Therefore, the speedup will be $= \dfrac{16000}{4060} = 3.941$.

Multiple Pipeline
Multiple pipeline architecture can be defined as a type of parallel architecture, formed using more than single independent pipeline in parallel. Thus, this architecture is a combination of pipeline and MIMD architectures.

Multiple SIMD
Multiple SIMD is a specific type of MIMD-SIMD architecture. More precisely, it can be defined as an MIMD type connection of a number of independent SIMD architectures. There are a number of control units for these architectures, each of which controls a subset of the PEs.

Dataflow Architecture
Another novel parallel architecture is the dataflow model. In this case, the program is represented by a graph of data dependencies as shown in Figure 2.34. The graph is mapped over a number of processors each of which is assigned an operation to be performed and the address of each node that needs the result. A processor performs an operation whenever its data operands are available. The operation of dataflow architectures is quite simple and resembles circular pipelining. A processor receives a message comprising data and the address of its destination node. The address is compared against those in a matching store. If the address is present, the matching address is extracted and the instruction is issued for execution. If not, the address is placed in the store for its partner to arrive. When the result is computed, a new message or token containing the result is sent to each of the destinations mentioned in the instructions (Culler *et al.*, 1999).

Systolic and Wavefront Arrays
Systolic arrays comprise SIMD, MIMD and pipeline architectures. They are driven by a single clock and hence behave like SIMD architectures. However, they differ from SIMD in that each PE has the option to do different operations. The individual array elements, on the other hand, are MIMD processors and pipeline computations take place along all array dimensions. The systolic array also differs from conventional pipelined function units in that the array structure could be non-linear, the pathways between PEs may be multidirectional and each PE may have a small amount of local instruction and data memory (Culler *et al.*, 1999). Replacing the central clock of the systolic arrays with the concept of data flow forms wavefront arrays and hence wavefront arrays can be regarded as an extension of systolic arrays.

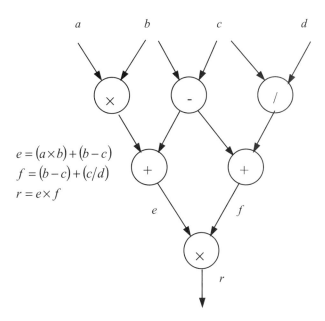

Figure 2.34. Data-flow graph

Single-program Multiple-data Architecture
The single-program multiple-data (SPMD) architecture combines the ease of SIMD programming with MIMD flexibility. This system is controlled by a single program and hence the name SPMD.

2.4 Summary

A large number of diverse types of parallel architectures are used worldwide. There is also no doubt that there are many other types in the research and/or development stage. However, not all of these suit particular applications. Thus, it is necessary to figure out which parallel architecture would be appropriate for what types of applications. This has essentially been the motivation for the classification of parallel architectures. Flynn first classified parallel architectures based on the instruction and data streams. His classification gives a broad picture of parallel architectures and all parallel architectures could be classified in terms of this broad classification principle. However, this is not enough to fully distinguish one parallel architecture from another and as a result further classifications in terms of more distinctive features have evolved. Such features include memory arrangements, interconnection mechanisms, communication between PEs, memory access time, nature of the processors incorporated in an architecture and so on. For example, when all the processors in an architecture share a single memory, it is called shared memory architecture whereas when each processor uses its own local memory it is called a distributed memory architecture. A number of architectures have evolved based on the interconnection mechanisms and arrangement of the PEs in the architecture. Homogeneous and heterogeneous types of parallel architecture evolved based on such issues, where a heterogeneous architecture comprises processors with different characteristics and a homogeneous architecture comprises processors with similar characteristics.

2.5 Exercises

1. Indicate the major classes of parallel architectures? Describe Flynn's classification of computers.

2. Distinguish between shared memory and distributed memory architectures.

3. What do you understand by UMA and NUMA as used in parallel architectures?

4. Indicate various characteristics of message passing architectures. How does message passing occur in a message passing architecture?

5. Classify parallel architectures on the basis of interconnection networks. Distinguish between static and dynamic interconnection networks.

6. Draw a comparative outline of various interconnection networks.

7. Describe the function of switches used in dynamic interconnection architectures. Briefly explain the working mechanism of a switch.

8. Distinguish between CISC and RISC architectures.

9. Explain the distinctive features of homogeneous and heterogeneous parallel architectures.

10. Indicate the characteristics of vector processor, array processor and DSP devices.

11. What do you understand by pipeline mechanism? Describe the working mechanism of a pipeline architecture.

12. Consider a four-segment pipeline processor executing 4000 tasks. Assume that the time it takes to process a sub-operation in each segment is equal to $30\,\text{ns}$. Determine the speedup for the pipeline system.

13. Consider the time delay of the five segments in the pipeline of Figure 2.33 as: $t_1 = 45\,\text{ns}$, $t_2 = 30\,\text{ns}$, $t_3 = 95\,\text{ns}$, $t_4 = 50\,\text{ns}$ and $t_5 = 70\,\text{ns}$. The delay

time of interface registers is $t_r = 15\,\text{ns}$. (a) How long should it take to add 100 pairs of numbers in the pipeline? (b) How can you reduce the total time to about half of the time obtained in part (a)?

14. Illustrate that the speedup of a four-segment floating-point adder for a large number of tasks is nearly 4.

15. How is a systolic array formed? Describe the features of a systolic array and warfront computers.

16. Describe the basic working principles of data-flow architecture.

3. Performance Evaluation Issues in Real-time Computing

3.1 Objectives

- To explore the need for performance evaluation.
- To focus on the issues that impact computing performance.
- To demonstrate the practical aspects of performance evaluation issues.

3.2 Introduction

Parallel processing has emerged as a key enabling technology in modern computing to meet the ever-increasing demand for higher performance, lower costs and sustained productivity in real-life applications. Despite the vastly increased computing power that is now available, there can still be limitations on the computing capability of digital processors in real-time signal processing and control applications for two reasons:

a. sample times have become shorter as greater performance demands are imposed on the system;

b. algorithms are becoming more complex as the development of signal processing and control theory leads to an understanding of methods for optimising system performance.

To satisfy these high-performance demands, microprocessor technology has developed at a rapid pace in recent years. This is based on

- Processing speed.
- Processing ability.
- Communication ability.
- Control ability.

Every year brings new devices, new functions, and new possibilities. An imaginative and effective architecture for today could be reality for tomorrow, and likewise, an ambitious proposal for today may be ideal for tomorrow. There are no absolute rules that one architecture is better than another. In terms of design strategy, performance and computing resources, every microprocessor possesses its own merits (Carr, 1990; Stone, 1990). However, many demanding complex signal processing and control algorithms cannot be satisfactorily realised with conventional uni-processor and multiprocessor systems. Alternative strategies where multiprocessor based systems are employed, utilising high performance processors and parallel processing techniques, could provide suitable methodologies (Ching and Wu, 1989; Jones, 1989; Tokhi *et al.*, 1992).

It is estimated that many modern supercomputers and parallel processors deliver only 10% or even less of their peak performance potential in a variety of applications (Moldovan, 1993). The causes of performance degradation are many. For instance, performance losses occur because of mismatches between algorithms and architectures. The degree of parallelism is the number of independent operations that may be performed in parallel. In complex signal processing and control systems, mismatches occur between software modules or hardware modules. For example, the communication network bandwidth may not correspond to the processor speed or that of the memory. Some of the parallelism is lost when an algorithm is actually implemented as a program, because of the inability of many languages to express parallelism. This may sometimes be recovered by using parallelising compilers. In addition, parallelism is lost when a program is executed on a particular parallel computing domain. The operating system that schedules and synchonises tasks, manages memory and other resources, impose restrictions that cause degradation of parallelism.

On the other hand, the number of available processors in an architecture may not match the size of the problem. The processor granularity may be larger than required and thus may waste resources, or the bandwidth and topology of an interconnection network may not correspond to algorithm requirements. All these mismatches contribute to real-time parallel performance degradation. Additionally, the scheduling and mapping of an algorithm onto a parallel computer is a very important and difficult task, because of disparities in the complexity of algorithms. Any small change in problem size when using different algorithms or different applications may have undesirable effects and can lead to performance degradation in real-time implementation (Hossain, 1995).

There will always be applications that are satisfied by a uniprocessor implementation. Before adopting a parallel processing solution, it must be clear that it possesses features that cannot be provided by a single processor. Investigations and practical experiences have revealed a number of features that play an important role in the performance of an architecture. The key components of performance evaluation are described in the following sections.

3.3 Performance Evaluation Issues

There are various factors that play important roles in the performance of an architecture. The main issues that impact on the performance evaluation criteria are:

    i.   hardware;
   ii.   algorithm;
  iii.   software;
  iv.   cost considerations.

These are shown in a chart in Figure 3.1 with further levels of detail. Some of these key issues are described below.

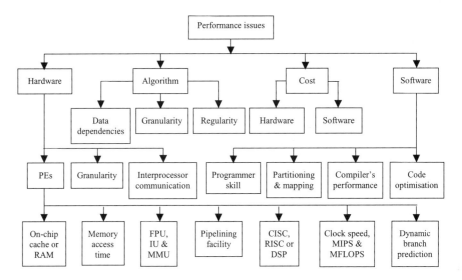

Figure 3.1. Main factors influencing performance of architectures

3.3.1 Hardware

Processing Elements
For high performance in sequential and parallel processing, with widely different architectures, performance measurements such as MIPS (million instructions per second), MOPS (million operations per second) and MFLOPS (million floating-point operations per second) are meaningless. Of more importance is to rate the performance of a processor on the type of program likely to be encountered in a particular application (Tokhi and Hossain, 1995).

    The different processors and their different clock rates, memory cycle times etc. all confuse the issue of attempting to rate the processors. In particular, there is an inherent difficulty in selecting processors in signal processing and control applications. As will be demonstrated in later chapters of the book, the ideal performance of a processor demands a perfect match between processor capability and program behaviour. Processor capability can be enhanced with better hardware technology, innovative architectural features and efficient resource management. From the hardware point of view, current performance varies according to whether the processor possesses a pipeline facility, is microcode/hardwire operated, has an internal cache (unified or isolated) or internal RAM, has a built-in math co-processor, floating point unit etc. Program behaviour, on the other hand, is difficult to predict due to its heavy dependence on application and runtime conditions. Other factors affecting program behaviour include algorithm design, data structure, language efficiency, programmer skill and compiler technology (Anderson, 1991; Hwang 1993). This chapter presents such issues within the framework of signal processing and control applications.

Digital signal processing devices are designed in hardware to perform concurrent add and multiply instructions and execute irregular algorithms efficiently, typically finite-impulse response (FIR) and infinite-impulse response (IIR) filter algorithms. Vector processors are designed to efficiently process regular algorithms involving matrix manipulations. However, many demanding complex signal processing and control algorithms cannot be satisfactorily realised with conventional computing methods. Alternative strategies where high-performance computing methods are employed could provide suitable solutions in such applications (Tokhi and Hossain, 1996).

Interprocessor Communication

For parallel processing, interprocessor communication between processing elements is an important issue used to compare the real-time performance of a number of parallel architectures and the suitability of an algorithm. The amount of data, the frequency with which the data is transmitted, the speed of data transmission, latency, and the data transmission route all significantly affect the intercommunication within the architecture. The first two factors depend on the algorithm itself and how well it has been partitioned. The remaining two factors are functions of the hardware. These depend on the interconnection strategy, whether tightly coupled or loosely coupled. Any evaluation of the performance of the interconnection must be, to a certain extent, quantitative. However, once a few candidate networks have been tentatively selected, detailed (and expensive) evaluation including simulation can be carried out and the best network selected for a proposed application (Agrawal *et al.*, 1986; Hossain, 1995).

In practice, the most common interprocessor communication techniques for general-purpose and special-purpose parallel computing are as shown in Figure 3.2. These are described below.

Shared memory communication. This is most widely used interprocessor communication method. Most commercial parallel computers utilise this type of communication method due to its simplicity, in terms of hardware and software. In addition, as most general-purpose microprocessors (CISC or RISC) do not have a serial or parallel communication link, these devices can communicate with each other only via shared memory. However, this is one of the slowest interprocessor communication techniques. A major drawback of this technique is how to handle reading and writing into the shared memory. In particular, while one processor (say, P1), reads/writes into the shared memory, the other processor (say, P2) has to wait until P1 finishes its read/write job. This mechanism causes an extremely high communication overhead. Performance degradation increases with higher data and control dependencies due to the higher communication overhead.

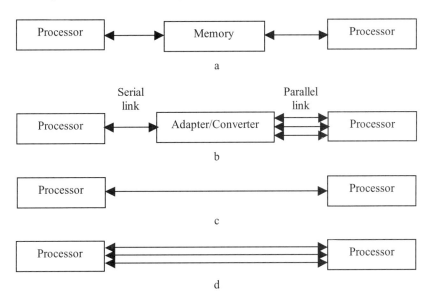

Figure 3.2. Interconnection strategies: a. shared memory communication; b. serial to parallel link communication ore vice versa; c. serial to serial link communication; d. parallel to parallel link communication

Serial to parallel communication or vice versa. This is a special type of interprocessor communication technique. It is most commonly used in heterogeneous architectures, where different processors possess different types of communication links. However, higher communication overhead is the major disadvantage of this type of interprocessor communication method. In this case, a serial to parallel or parallel to serial conversion is required to communicate between the processors. In practice, this conversion takes a significant amount of time, which in turn affects the real-time performance of the system.

Serial to serial communication. This type of interprocessor communication is used where processors possess serial link communication facility. This is particularly common in homogeneous parallel architectures that possess a serial link facility. Transputer-based homogeneous architectures commonly use this type of interprocessor communication method.

Parallel to parallel communication. Some special-purpose processors possess parallel links. Homogeneous architectures commonly use this type of link, where the processors are identical and possess parallel links. In practice, this is the best interprocessor communication method, with minimum communication overhead. Some special-purpose processors, including the Texas Instruments TMS320C40, and the Analogue Devices SHARC DSP processors possess this type of parallel link.

Hardware Granularity
This is an important issue, which plays a vital role in achieving higher performance in real-time parallel processing. Hardware granularity can be defined as the ratio of computational performance over the communication performance of each processor within the architecture. More clearly,

$$\text{Hardware Granularity} = \frac{\textit{Runtime length of a task}}{\textit{Communication overhead}} = \frac{R}{C}$$

When R/C is very low, it is unprofitable to use parallelism. When R/C is very high, parallelism is potentially profitable. A characteristic of fine-grain processors is that they have fast interprocessor communication, and can therefore tolerate small task sizes and still maintain a satisfactorily high R/C ratio. However, medium-grain or coarse-grain processors with slower interprocessor communication will produce correspondingly lower R/C ratios if their task sizes are also small.

In general, fine-grain architectures can perform algorithmic parallelism efficiently, while coarse-grain architectures are more suited to strategies such as functional parallelism, where task sizes are large and interprocessor communication is relatively infrequent.

3.3.2 Algorithms

There are three different problems to be considered in implementing algorithms on parallel processing systems:

   i.   identifying parallelism in the algorithm;
   ii. partitioning the algorithm into subtasks;
  iii. allocating the tasks to processors.

To identify algorithmic parallelism, it is essential to explore the basic structural features of algorithms that are dictated by their data and control dependencies. The presence of dependencies indicates complexity of an algorithm and, in turn, communication overhead in parallel processing. Different levels of dependencies can be distinguished, as:

- Block or coarse grain computation level.
- Statement or fine grain level.
- Variable level, and even.
- Bit level.

The study of dependencies between larger blocks of computation is more useful when applied to parallel architectures. At the other extreme, the study of dependencies at bit level is useful in designing efficient microprocessors, especially the ALUs.

These also include interprocessor communication (discussed in the earlier section), issues of granularity of the algorithm and of the hardware and regularity of the algorithm. Task granularity is similar to hardware granularity and can be defined as the ratio of computational demand over the communication demand of the task. Typically a high compute/communication ratio is desirable. The concept of task granularity can also be viewed in terms of compute time per task. When this is large, it is a coarse-grain task implementation. When it is small, it is a fine-grain task implementation. Although, large grains may ignore potential parallelism, partitioning a problem into the finest possible granularity does not necessarily lead to the fastest solution, as maximum parallelism also has maximum overhead, particularly due to increased communication requirements. Therefore, when partitioning the algorithm into subtasks and distributing these across processing elements, it is essential to choose an algorithm granularity that balances useful parallel computation against communication and other overheads (Hossain, 1995).

Regularity is a term used to describe the degree of uniformity in the execution thread of the computation. Many algorithms can be expressed by matrix computations. This leads to the so called regular iterative type of algorithms due to their very regular structure. In implementing these types of algorithms, a vector processor will be expected to perform better. Moreover, if a large amount of data is to be handled for computation in these types of algorithms, the performance will be further enhanced if the processor has more internal data cache, instruction cache and/or a built-in math coprocessor.

3.3.3 Software Issues

Software support is needed for the development of efficient programs in high-level languages. The ideal performance of a computer system demands a perfect match between machine capability and program behaviour. Program performance is based on turnaround time, which includes disk and memory accesses, input and output activities, compilation time, operating system overhead, and CPU time. In order to shorten the turnaround time, one can reduce all these time factors. Minimising the runtime memory management, using efficient partitioning and mapping of the program, and selecting an efficient compiler for specific computational demand could enhance the performance. Compilers have a significant impact on the performance of the system. This means that some high-level languages have advantages in certain computational domains, and some have advantages in other domains. The compiler itself is critical to the performance of the system as the mechanism used and efficiency obtained when taking a high-level description of the application and transforming it into a hardware dependent implementation differs from compiler to compiler (Bader and Gehrke, 1991; Tokhi *et al.*, 1995). Identifying the best compiler for the application in hand is, therefore, especially challenging.

Scheduling and mapping are also important issues to be considered in software development. There are two main approaches to allocating tasks to processors: statically and dynamically. In static allocation, the association of a group of tasks with a processor is resolved before runtime and remains fixed throughout the execution, whereas in dynamic allocation, tasks are allocated to processors at runtime according to certain criteria, such as processor availability, intertask dependencies and task priorities. Whatever method is used, a clear appreciation is required of the overheads and parallelism/communication trade-off as mentioned earlier. Dynamic allocation offers greater potential for optimum processor utilisation, but it also incurs a performance penalty due to scheduling overheads and increased communication requirements, which may prove unacceptable in some real-time applications.

Performance is also related to the program optimisation facility of the compiler, which may be machine dependent. The goal of program optimisation is, in general, to maximise the speed of code execution. This involves several factors such as minimisation of code length and memory accesses, exploitation of parallelism, elimination of dead code, in-line function expansion, loop unrolling and maximum utilisation of registers. The optimisation techniques include vectorisation using pipelined hardware and parallelisation using multiprocessors simultaneously (Hossain, 1995).

3.3.4 Cost Consideration

Despite the undoubted success of embedded digital controllers in industrial signal processing and control applications, an increasing number of parallel processors are appearing on the market. Increased computational speed is of course the primary benefit of parallel processing for real-time implementation. This allows faster systems to be controlled and gives the control engineer the choice of added complexity in the control algorithm. Easy expansion within a uniform hardware and software base is another feature of concurrent systems, since it is possible to add processors as required. This has important implications for reduced development and maintenance costs. Parallelism is beneficial when it successfully yields higher performance with reasonable hardware and software cost.

Problem 3.1: *Consider three identical processors executing a set of tasks T1 to T10. The execution times associated with the tasks are as in Table 3.1. Determine the total idle time of the three processors without data dependencies and with data dependencies according to the diagram shown in Figure 3.3.*

Solution: Task allocation for the three processors without dependencies and with dependencies are shown in Tables 3.2 and 3.3, where P1, P2 and P3, represents the three processors.

Table 3.1. Tasks and execution time for a single processor

Tasks	Execution time (s) for single processor
T_1	4
T_2	3
T_3	5
T_4	4
T_5	4
T_6	3
T_7	4
T_8	5
T_9	4
T_{10}	3

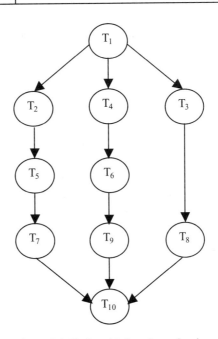

Figure 3.3. Tasks with data dependencies

It is noticed from the task allocations in Table 3.2 that there is no idle time for any of the three processors when there is no data dependencies among the tasks. On the other hand, with dependencies as shown in Figure 3.3, it is noted from the task allocation Table 3.3, that for

Processor P1, utilisation time is 15 s and idle time is 3 s
Processor P2, utilisation time is 11 s and idle time is 7 s
Processor P3, utilisation time is 13 s and idle time is 5 s

Hence, total idle time of the architecture due to data dependencies is 15 s, which implies that as compared to the total utilisation time (39 s), the idle time of the architecture is almost 38.5%.

Table 3.2. Task allocation without dependencies

	Execution time (s)												
	1	2	3	4	5	6	7	8	9	10	11	12	13
P1	T_1				T_3					T_4			
P2	T_2			T_5			T_6				T_{10}		
P3	T_7				T_8					T_9			

Table 3.3. Task allocation with dependencies based on Figure 3.3

	Execution time (s)													
	1	2	3	4	5	6	7	8	9	10	11	12	13	14
P1	T_1				T_2				T_5			T_7		
P2						T_4				T_6				T_9
P3					T_3					T_8				

	Execution time (s), continued			
	15	16	17	18
P1				
P2				
P3		T_{10}		

3.4 Case Study

To demonstrate the critical performance evaluation issues described earlier, a number of experiments have been conducted and the results are presented and discussed in this section. This will help the reader to develop a practical understanding of the concepts. Several algorithms are utilised in these experiments including a finite difference simulation of an aluminium type cantilever beam of length $L = 0.635\,\mathrm{m}$, mass $m = 0.037\,\mathrm{kg}$ and beam constant $\mu = 1.351$, and an LMS adaptive filter algorithm with rate $\eta = 0.04$. These algorithms are described further in Chapter 6. The hardware platforms utilised in these experiments incorporate the INMOS T805 (T8) transputer, Texas Instruments TMS320C40 (C40) DSP devices and the Intel 80i860 (i860) vector processor. These are described in Appendix B.

3.4.1 Interprocessor Communication

When several processors are required to work co-operatively on a single task, one expects frequent exchange of data among the subtasks that comprise the main task. To explore the real-time performance of parallel architectures, investigations into interprocessor communication have been carried out. The interprocessor communication techniques for different architectures are

- T8–T8: serial communication link.
- C40–C40: parallel communication link.
- T8–C40: serial to parallel communication link.
- T8–i860: shared memory communication.

The performance of these interprocessor communication links is evaluated by utilising a similar strategy for exactly the same data block without any computation during the communication time, i.e., blocking communications. It is important to note that, although the C40 has potentially high data rates, these are often not achieved when using the 3L Parallel C due to the routing of communications via the microkernel (which initialises a DMA channel). This, as noted later, incurs a significant setting up delay and becomes particularly prohibitive for small data packets.

To investigate interprocessor communication speed using the communication links indicated above, 4000 floating-point data elements were used. The communication time was measured as the total time to send data from one processor to another and receive it back. In the case of C40–T8 and C40–C40 communications, the speed of a single line of communication was also measured using bi-directional data transmission in each of the 4000 iterations. This was achieved by changing the direction of the link at every iteration when sending and receiving the data. Figure 3.4 shows the communication times for the various links, where, (1) represents a single bi-directional line of communication and (2) represents a pair of uni-directional lines of communication. Note that, as expected, the C40–C40 parallel pair of lines of communication performed fastest and the C40–T8 serial to parallel single line of communication the slowest of these communication links. Table 3.4 shows the relative communication times with respect to the C40–C40 parallel pair of lines of communication. It is noted that the C40–C40 parallel pair of lines of communication was 10 times faster than the T8–

T8 serial pair of lines of communication and nearly 15 times faster than the i860–T8 shared memory pair of lines of communication. The slower performance of the shared memory pair of lines of communication as compared to the T8–T8 serial pair of lines of communication is due to the extra time required to write data into and read data from the shared memory. The C40–T8 serial to parallel pair of lines of communication involves a process of transformation from serial to parallel, while transferring data from T8 to C40, and vice versa, when transferring data from C40 to T8, making the link about 17.56 times slower than the C40–C40 parallel pair of lines of communication. As noted, the C40–C40 parallel single line of communication performed about 94 times slower than the C40–C40 parallel pair of lines of communication. This is due to the utilisation of a single bi-directional line of communication in which, in addition to the sequential nature of the process of sending and receiving data, extra time is required to alter the direction of the link (dataflow). Moreover, there is a setting-up delay for each communication performed. This is of the order ten times that of the actual transmission time. These aspects are also involved in the C40–T8 serial to parallel single line of communication which performed 115.556 times slower than the C40–C40 parallel pair of lines of communication due to the extra time required for the process of transformation from serial to parallel and vice versa.

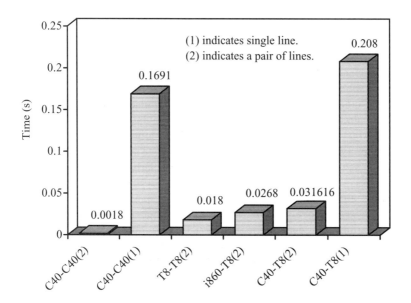

Figure 3.4. Interprocessor communication times for various links

To investigate the interprocessor communication issue further in a parallel processing platform the beam simulation algorithm was considered and implemented on a homogeneous transputer architecture. It was found through numerical simulations that for the purpose of this investigation reasonable accuracy in representing the first few (dominant) modes of vibration is achieved by dividing the beam into 19 segments. Moreover, a sample period of $\Delta t = 0.3$ ms, which is sufficient to cover all the resonance modes of vibration of the beam of interest, was selected. In this investigation, the total execution times achieved by the architectures in implementing the simulation algorithm over 20,000 iterations was considered in comparison to the required real time (calculated as the product of the sampling time Δt and total number of iterations).

Table 3.4. Interprocessor communication times with various links (LP1) relative to the C40–C40 parallel pair of lines of communication (LP2)

Link	T8–T8(2)	C40–C40(1)	C40–T8(2)	C40–T8(1)	i860–T8(2)
LP1/LP2	10.00	93.889	17.5644	115.556	14.889

The algorithm considered consists of computations of the deflection of 19 equal-length segments. The computation for each segment requires information from two previous and two forward segments. Figure 3.5 shows the required communication and logical distribution of segments among networks comprising two, three and six PEs. In practice, for the three-PE architecture, the load cannot be equally distributed due to communication overheads. PE2 used for the computation of segments 7, 8, 9, 10, 11 and 12 requires more communication time than the other two PEs. It was found through this investigation that the communication time for PE2 was nearly equivalent to the computation time for 2.5 segments. This led to allocating seven segments (1,2,3,4,5,6,7) to PE1, five segments (8,9,10,11,12) to PE2 and the remaining seven segments to PE3 to obtain optimum performance. A similar situation occurs in implementations using more than three PEs. Here, it is essential to note that if the number of PEs is more than three, there will be additional communication overheads occurring in parallel with others. As a result of this the communication overhead may be bounded within certain limits. Another important factor for reducing communication overhead would be to pass messages in a block of two segments from one processor to another, rather than sending one segment at a time. Using such a strategy, the algorithm was implemented on networks of up to nine T8s.

To explore the computation time and the computation with communication overhead, the performance of the architecture was investigated by breaking the algorithm into fine grains (one segment as one grain). Considering the computation for one segment as a base, grains were then computed on a single T8 increasing the number of grains from one to nineteen. The theoretical linear computation time of one to nineteen segments and the actual computation time are shown in Figure 3.6. Note that the theoretical computation time is more than the actual computation time. This implies the RISC nature of the T8 processor. The actual computation time for a single T8 was then utilised to obtain the actual computation time for the multiprocessor system without communication overhead. Figure 3.7 shows the real-time performance (i.e., computation with communication overhead) and the actual computation time with one to nine PEs. The difference between the real-time performance and the actual computation time is the communication overhead, shown in Figure 3.8. It is noted in Figures 3.7 and 3.8 that, due to communication overheads, the computing performance does not increase linearly with increasing number of PEs. The performance remains nearly flat with a network of more than six PEs. Note that the increase in communication overhead at the beginning, with less than three PEs, is more pronounced and remains nearly at the same level with more than five PEs. This is due to the communication overheads among the PEs, which occur in parallel.

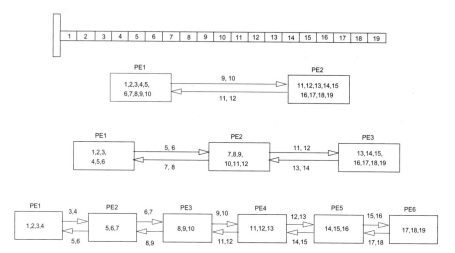

Figure 3.5. Task allocation and communication of two, three and six processors

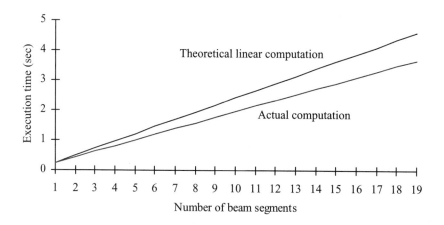

Figure 3.6. Execution time for flexible beam segments on a single T8

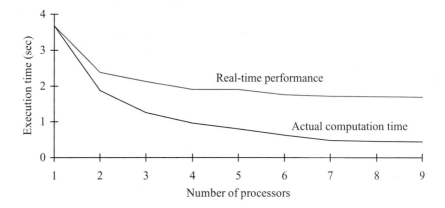

Figure 3.7. Execution time of the simulation algorithm running on the transputer network

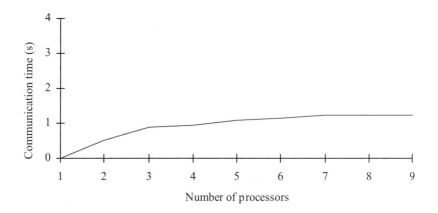

Figure 3.8. Communication overhead of the transputer network

3.4.2 Compiler Efficiency

In this section results of investigations of the performance evaluation of several compilers are presented and discussed. The compilers involved are the 3L Parallel C version 2.1, Inmos ANSI C and Occam (see Appendix B for details). All these compilers can be utilised with the computing platforms considered in this investigation. It has previously been reported that, although Occam is a more hardware oriented and straightforward programming language for parallel processing, it may not be as suitable as the Parallel C or ANSI C compilers for numerical computations (Bader and Gehrke, 1991). To obtain a comparative performance evaluation of these compilers, the flexible beam simulation algorithm was coded, for 19 equal-length beam sections, into the three programming languages and run on a T8. Figure 3.9 shows the execution times to implement the simulation algorithm, over 20,000 iterations, using the three compilers. It is noted that the performances with Parallel C and ANSI C are nearly at a similar level and about 1.5 times faster than the Occam. This was further investigated with a linear algebraic equation. Table 3.5 shows the performances with integer and floating-point operations with and without an array, where

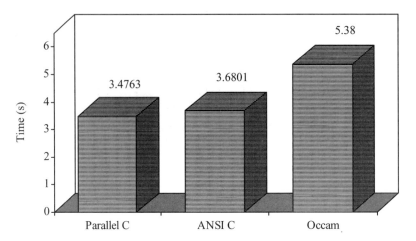

Figure 3.9. Performance of the compilers when implementing the simulation algorithm on the T8 transputer

Table 3.5. Comparison of compilers for different types of data processing

Compiler	Floating type data processing		Integer type data processing	
	With array 20000/40000	Without array 20000/40000	With array 20000/40000	Without array 20000/40000
3L Parallel C	0.1327 / 0.2488	0.1263 / 0.2333	0.1327 / 0.2488	0.1263 / 0.2333
ANSI C	0.1328 / 0.2444	0.0127 / 0.0227	0.1328 / 0.2444	0.0126 / 0.0226
Occam	0.1078 / 0.2052	0.1078 / 0.2044	0.1960 / 0.3825	0.1905 / 0.3698

The code without an array declares: $z = (x + i * y - x * i)/(x * x + y * y)$

The code with an array declares:
$$z(i) = [x(i) + i * y(i) - x(i) * i]/[x(i) * x(i) + y(i) * y(i)]$$

with $i = 0, 1, ..., 20,000$, integer $x = 55$, $y = 25$ and floating $x = 55.02562$, $y = 25.13455$. It is noted that better performance is achieved throughout with the ANSI C compiler than the Occam, except in a situation where the computation involved is floating type data processing with a declaring array. Compared to ANSI

C, better performance is achieved with Parallel C for both integer and floating type computation with an array. Better performance is achieved with the Occam compiler for floating type computation than with Parallel C. It is also noted that for double data handling 1.9 times more execution time was required with Occam. In contrast, 1.87 times more execution time was required with ANSI C and Parallel C. This implies that for large amounts of data handling, the runtime memory management problem can be solved with Parallel C and ANSI C more efficiently than with the Occam compiler.

3.4.3 Code Optimisation

The code optimisation facility of compilers for the selected hardware is another important component affecting the real-time performance of a processor. Optimisation facilities almost always enhance the real-time performance of a processor. The i860 and the C40 have many optimisation features (Portland Group Inc., 1991; Texas Instruments, 1991a,b). The TMS320 floating-point DSP optimising C compiler is the TMS320 version of the 3L Parallel C compiler (Texas Instruments, 1991c). It has many options, constituting three levels of optimisation, which aid the successful optimisation of C source code files on the C40. The Portland Group (PG) C compiler is an optimising compiler for the i860 (Portland Group Inc., 1991). It incorporates four levels of optimisation.

To measure the performance attainable from the compiler optimisers, so as to fully utilise the available features, experiments were conducted to compile and run the LMS and beam simulation algorithms on the i860 and the C40. To study the effect of the PG optimising compiler, the LMS algorithm was compiled with the number of weights set to five and $\eta = 0.04$. The algorithm was implemented on the i860 with four levels of optimisation and the execution time of the processor to implement the algorithm over 1000 iterations was recorded. Similarly, the beam simulation algorithm was compiled and implemented on the i860 with five beam segments and $\Delta t = 0.055$ ms. The execution time of the processor to implement the algorithm over 20,000 iterations was recorded with each of the four levels of optimisation. Figure 3.10 shows the execution times achieved when implementing the LMS and the beam simulation algorithms, where level 0 corresponds to no optimisation. The corresponding execution time speedups achieved with each optimisation level when implementing the LMS and beam simulation algorithms are shown in Figure 3.11. It is noted in Figures 3.10 and 3.11 that the performance of the processor when implementing the LMS algorithm is enhanced significantly by higher levels of optimisation. Enhancement in the case of the beam simulation algorithm, on the other hand, is not significant beyond the first level. The disparity in the speedups of the two algorithms is thought to be due to the type of operations performed by the optimiser. As the LMS algorithm has multiple nested loops, the compiler is able to recognise the structures involved and restructure the actual code so that it is still functionally equivalent but suits the CPU architecture better. Branches to subroutines, for instance, are in-lined, which cuts out procedure call overheads, hence reducing the execution time. The beam simulation algorithm, however, is already in matrix format and thus does not have as much room for improvement (Portland Group Inc., 1991).

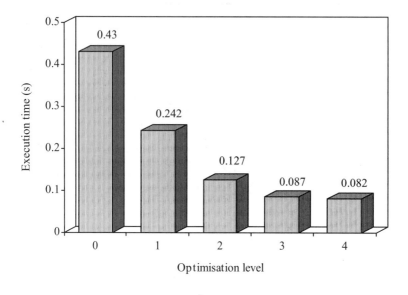

a

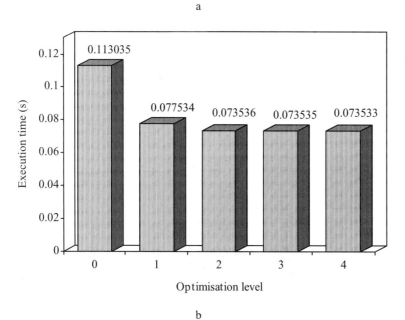

b

Figure 3.10. Execution times of the i860 when implementing the algorithms with the Portland Group C compiler optimiser: a. LMS algorithm; b. beam simulation algorithm

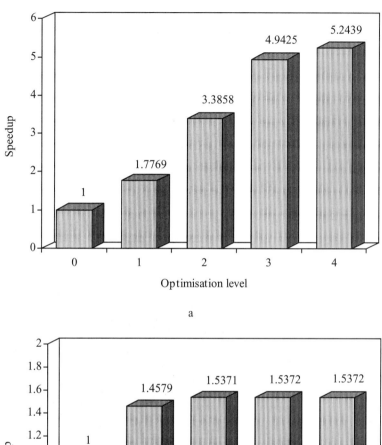

a

b

Figure 3.11. Speedup with the Portland Group C compiler optimiser when implementing the algorithms on the i860: a. LMS algorithm; b. beam simulation algorithm

To study the effect of the optimisers further on the performance of the system, optimisation level 0 (no optimisation) and level 2 were used with the 3L parallel C optimiser to implement the algorithms on the C40. Similarly, with the i860, using the PG compiler, optimisation level 0 and level 4 were utilised. The LMS and beam simulation algorithms were coded for various task sizes by changing the number of weights in the case of the LMS algorithm and the number of segments in the case of the beam simulation algorithm. The algorithms were implemented on the i860 and the C40.

Figures 3.12 and 3.13 show the execution times achieved by the processors when implementing the LMS and beam simulation algorithms over 1000 and 20,000 iterations respectively. It is noted that the execution time in each case varies approximately linearly with the task size. With the LMS algorithm, as noted, the relation has a relatively smaller gradient for less than 10 and 20 weights with the i860 and the C40 respectively. For an increased number of weights, the gradient is larger. Such a phenomenon is most likely associated with the dynamic memory management conditions of the processor in each case. In the case of the beam simulation algorithm, as noted in Figure 3.13, such a situation is clearly not evident. This suggests that with the amount of data handling involved in implementing this algorithm both processors appear to have required utilisation of the lower level memory throughout. The slight variation in gradient noted in Figure 3.13(b) with the algorithm implemented on the C40 is most likely due to computational error.

It is noted in Figures 3.12 and 3.13 that the enhancement in performance of the processors when implementing the LMS algorithm with optimisation is substantially greater than when implementing the beam simulation algorithm. This, as discussed above, is due to the structure of the algorithms where, for the LMS algorithm, the features of the optimisation are well exploited but not for the beam simulation algorithm. This is further evidenced in Figure 3.14 showing the corresponding speedups achieved by optimisation when implementing the algorithms on the i860 and the C40. It is important to note that optimisation with the C40 offers significant enhancement when implementing the LMS algorithm. However, enhancement occurs within a specific bandwidth of the task size (number of filter weights). A similar trend would also be expected with the speedup achieved in the case of the i860 when implementing the LMS algorithm. However, due to the better data handling capability of the i860 the upper roll-off point is expected to occur for a larger task size in comparison to that with the C40 implementation.

The execution time speedups achieved with code optimisation when implementing the beam simulation algorithm on the processors, as noted in Figure 3.14(b), reach a similar level with both the i860 and the C40. At small task sizes the speedup with the C40 is relatively larger and continues to decrease with increase in the task size. The speedup with implementation on the i860, on the other hand, increases with task size rapidly at the lower end and then more slowly beyond 20 beam segments. This suggests that the optimisation for the i860 performs better than that for the C40 in this type of application.

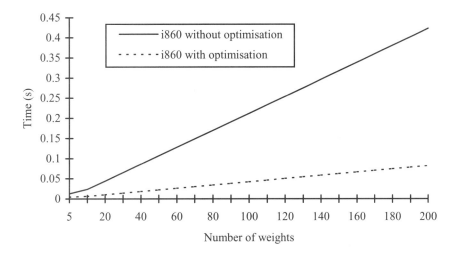

a

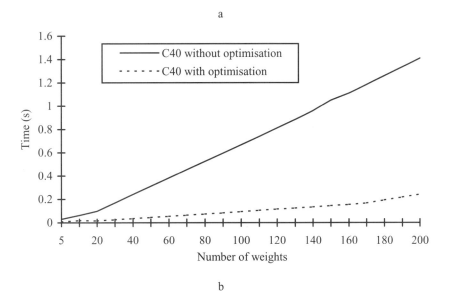

b

Figure 3.12. Execution times of the processors when implementing the LMS algorithm: a. with the i860; b. with the C40

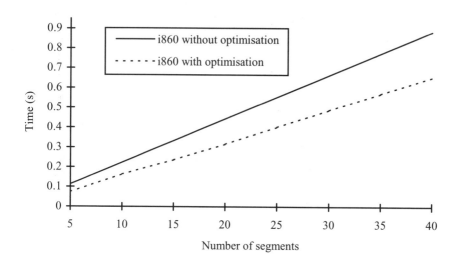

a

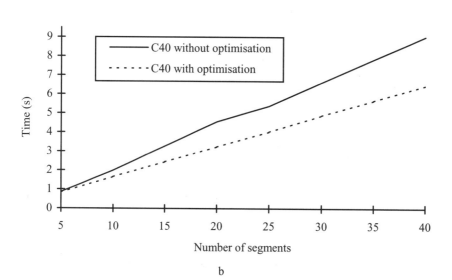

b

Figure 3.13. Execution times of the processors when implementing the beam simulation algorithm: a. with the i860; b. with the C40

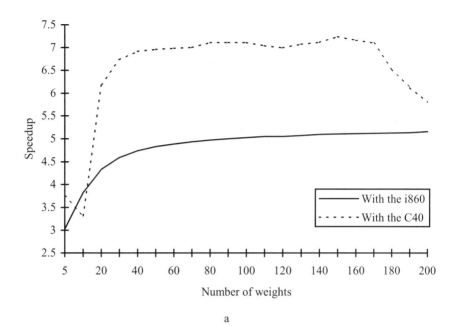

a

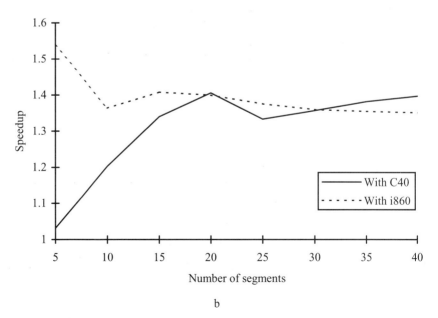

b

Figure 3.14. Optimisation speedup with the processors when implementing the algorithms: a. LMS algorithm; b. beam simulation algorithm

3.5 Summary

The important issues relevant to the performance evaluation of parallel computing have been described and demonstrated with a number of case studies. The need for performance evaluation, and issues that play vital roles in performance enhancement of parallel computing have also been highlighted. The performance evolution of processors in relation to task size, compiler efficiency for numerical computation and code optimisation have been investigated. A set of case studies has been provided to demonstrate the practical aspects of the issues described.

3.6 Exercises

1. What are the needs for parallel processing? Does parallel processing always enhance computing performance?

2. Distinguish between static and dynamic task allocation. Find the idle time of a homogeneous architecture of four processors executing the tasks of the algorithm given in Table 3.6.

3. Draw a chart showing the main factors influencing the performance of parallel processing. Among these, discuss the software issues in real-time computing.

4. Discuss the key components that play vital roles in real-time performance evaluation. Find the idle time of a homogeneous architecture of five processors executing the tasks of the algorithm given in Table 3.6, with the data dependencies of the tasks as shown in Figure 3.15.

Table 3.6. Tasks and execution time for single processor

Tasks	Execution time (s) for single processor
T_1	4
T_2	3
T_3	5
T_4	4
T_5	4
T_6	3
T_7	4
T_8	5
T_9	4
T_{10}	4
T_{11}	3
T_{12}	3

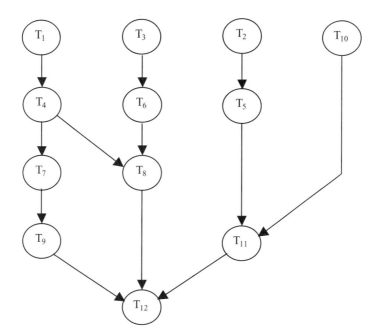

Fig. 3.15. Data dependencies of the tasks

4. Performance Metrics

4.1 Objectives

- To introduce techniques for performance measurement.
- To describe performance metrics.
- To demonstrate the practical significance of performance metrics through a case study.

4.2 Introduction

One of the popular methods used to characterise the performance of applications and systems is benchmarking. Benchmarks are used to measure and to predict the performance of computer systems. Moreover, benchmarking is used to reveal the architectural weaknesses and stengths. In general, a benchmark can be a full-fledged application or just a kernel, which is a much smaller and simpler program extracted from the application while maintaining the main characteristics. A benchmark can be a program that does real work or a synthetic program specifically designed for benchmarking. Benchmarks can be classified according to application, such as scientific computing, commercial applications, network services, multimedia applications, and signal processing. Common benchmarks for parallel processing are the NAS (Numerical Aerodynamic Simulation) program developed by NASA, Splash from Stanford University and PARKBENCH (PARallel Kernels and BENCHmarks) developed by a group of people interested in parallel computer benchmarking. The most popular mixed (both sequential and parallel computing) benchmarking family is SPEC (Standard Performance Evaluation Corporation) (Moldovan, 1993).

A parallel algorithm that solves a problem well using a fixed number of processors with a particular architecture may perform poorly if either of these parameters change. Analysing the performance of a given parallel algorithm/architecture calls for a comprehensive method that accounts for scalability: the system's ability to increase speedup as the number of processors increases. Scalability has been used in practice to describe the demand for proportionate changes in performance with adjustments in system size. Thus, scalability can intuitively be defined as a property that exhibits performance linearly proportional to the number of processors employed (Sun and Rover, 1994). Several parallel performance metrics that measure scalability have been proposed (Grama *et al.*, 1993; Gustafson, 1988, 1992; Sun and Ni, 1993; Sun and Rover, 1994; Worley, 1990).

Speedup is one of the most commonly used metrics for parallel processing. There are three known notions of speedup: fixed-size speedup, fixed-time speedup and memory-bounded speedup (Sun and Ni, 1993; Sun and Rover, 1994). Fixed-size speedup fixes the problem size and emphasises how fast a problem can be solved. Fixed-size speedup does not increase linearly with the number of processors; instead, it tends to saturate. This is true of all parallel systems, and is often referred to as Amdahl's law (discussed in Section 4.4.1). Fixed-size speedup is bounded by the reciprocal of the serial fraction of the algorithm. This hard performance limitation implies that fixed-size speedup is inadequate (Nussbaum and Agrawal, 1991). Fixed-time speedup argues that parallel computers are designed for otherwise intractably large problems. It fixes the execution time and emphasises how much more work can be done with parallel processing within the same time. Memory-bounded speedup assumes that the memory capacity, as a physical limitation of the machine, is the primary constraint on large problem sizes. It allows memory capacity to increase linearly with the number of processors. Both fixed-time and memory-bounded speedups are forms of scaled speedups. Each allows the problem size to increase with the system size. The difference between the two speedups is that fixed-time speedup uses execution time to limit the problem size whereas memory-bounded speedup uses memory capacity to limit the problem size. The term scaled speedup is used for memory-bounded speedup by many authors (Gustafson, 1988; Nussbaum and Agrawal, 1991).

Grama and co-workers have proposed the concept of isoefficiency as a measure of scalability of parallel algorithms (Grama *et al.*, 1993). Isoefficiency fixes the efficiency and measures the amount by which the work must be increased to keep the efficiency unchanged. An isoefficiency function, $f(N)$, where N is the number of processors, is defined as the amount of work needed to maintain the efficiency. The value of $f(N)$ could be arbitrarily large. Efficiency is defined as speedup divided by N. So, constant efficiency means that speedup increases linearly with system size. Thus, the concept of isoefficiency still uses speedup as the performance metric. The significant improvement in this concept is that it is not dependent on any of the notions of the three speedup models. By using an isoefficiency function, the problem size is allowed to increase without bound to attain the requisite efficiency.

Nussbaum and Agrawal (1991) proposed a definition of scalability of parallel machines. Accordingly, the scalability of a parallel machine, $\Psi(W)$, where W is the problem size, is defined as the best speedup of the architecture over the best speedup of an ideal parallel machine. This definition of $\Psi(W)$ still relies on speedup as the performance metric. However, it measures the speedup differently;

it determines the best speedup over the number of processors used, given an unbounded number of processors.

When speed is the goal, the power to solve problems of some magnitude in a reasonably short period of time is sought. Speed is a quantity that ideally would increase linearly with system size. Thus, it is consistent with the scalability property. Based on this reasoning, Sun and Rover (1994) proposed the isospeed approach. This is described by the average unit speed i.e., the speed achieved by a given computing system divided by the number of processors N. The average unit speed is referred to as average speed when the context is clear. Saying that the speed of a computing system is linearly proportional to the system size is the same as saying that the average speed is constant, independent of system size. Using average speed and the above intuitive definition of scalability a definition for a scalable algorithm–machine combination in which the parallel machine is homogeneous can be given as one in which the average speed of the algorithm on the given machine remains constant with increasing numbers of processors, provided the problem size can be increased proportionally (Sun and Rover, 1994). Scalability is thus expressed in terms of system size. In general, increasing the problem size will increase the computation/overhead ratio, and therefore, increase the speed. This is especially true for parallel processing where the computation-to-communication ratio increases with problem size for most algorithms. For a large class of algorithm–machine combinations, the average speed can be maintained by increasing the problem size. The necessary problem size increase varies with algorithms, machines and their combination. This variation provides a quantitative measurement for scalability.

Due to the substantial variation in computing capabilities of PEs, the traditional parallel performance metrics of homogeneous architectures are not suitable for heterogeneous architectures in their current form. Note, for example, that speed and efficiency provide measures of performance of parallel computation relative to sequential computation on a single processing node. In this manner, the processing node is used as a reference node. In a heterogeneous architecture, such a reference node representing the characteristics of all PEs, is not readily apparent. In this chapter, such a reference node is identified by introducing the concept of virtual processor. Moreover, it is argued that a homogeneous architecture can be considered as a subclass of heterogeneous architectures. In this manner, the performance metrics developed for heterogeneous architectures will be general and applicable to both classes of architectures.

Attempts have previously been made to define speedup of a heterogeneous architecture as the ratio of minimum sequential execution time among the PEs over the parallel execution time of the architecture (Yan et al., 1996; Zhang and Yan, 1995). In this manner, the best PE in the architecture is utilised as the reference node and the efficiency of the architecture is defined accordingly. Although it has been shown that this definition, in specific situations, transforms to that of a homogeneous architecture, such a transformation does not hold in general. Moreover, the concept does not fully exploit the capabilities of all the PEs in the architecture; instead, it relies on the performance of the best PE. The concept introduced in this chapter ensures that the full capabilities of the PEs are exploited by maximising the efficiency of the architecture.

4.3 Sequential Processing

It has previously been reported that in a large number of applications, the performance of a processor, in terms of execution time, in implementing an application algorithm generally evolves linearly with the task size (Tokhi *et al.*, 1996). With some processors, however, anomalies in the form of a change of gradient (slope) of execution time to task size are observed. These are mainly due to runtime memory management conditions of the processor where, up to a certain task size the processor may find the available cache sufficient, but beyond this it may require to access lower level memory. Despite this, the variation in slope is relatively small and the execution time to task size relationship can be considered linear. Alternatively, and if the variation is large, the relationship can be considered as piece-wise linear. In either case, this means that a quantitative measure of performance of a processor in a given application can be given adequately by the average ratio of task size to execution time or the average speed. Alternatively, the performance of the processor can be measured as the average ratio of execution time per unit task size, or the average (execution time) gradient. In this manner, a generalised performance measure of a processor relative to another in a given application can be obtained.

Let the average speeds with two processors p_1 and p_2 in an application, over a range of task sizes, be denoted by V_1 and V_2 respectively. The generalised sequential (execution time) speedup $S_{1/2}$ of p_1 relative to p_2 when implementing the application algorithm can thus be defined as

$$S_{1/2} = \frac{V_1}{V_2} \tag{4.1}$$

Alternatively, if the corresponding average gradients with p_1 and p_2 for the application, over a range of task sizes, are given by G_1 and G_2 respectively, $S_{1/2}$ can be expressed as

$$S_{1/2} = \frac{G_2}{G_1} \tag{4.2}$$

The concept of generalised sequential speedup described above can also be utilised to obtain a comparative performance evaluation of a processor for an application under various processing conditions, for example with and without code optimisation.

The concept of speed, assumed to be constant for a processor in a given application, has previously been utilised to derive an expression for the generalised speedup as the ratio of parallel speed (of a parallel architecture) over sequential speed (of a single processor) (Sun and Gustafson, 1991; Sun and Rover, 1994). The generalised speedup introduced above, however, reflects the relative performance of two uni-processor architectures in an application and the same processor under two different processing conditions.

4.4 Parallel Processing

The performance of a parallel architecture is commonly measured in terms of speedup and efficiency. The concepts of speedup and efficiency have been studied extensively for homogeneous architectures. However, little work has been reported on these concepts for heterogeneous architectures.

4.4.1 Homogeneous Architectures

For a homogeneous architecture, (fixed-load), parallel speedup (S_N) is defined as the ratio of the execution time (T_1) on a single processor, to the execution time (T_N) on N processors

$$S_N = \frac{T_1}{T_N} \tag{4.3}$$

The theoretical maximum speedup that can be achieved with a parallel architecture of N identical processors working concurrently on a problem is N. This is known as the ideal speedup. In practice, the speedup is much less, since some architectures do not perform to the ideal level due to conflicts over memory access, communication delays, inefficiency in the algorithm and mapping to exploit the natural concurrency in a computing problem (Hwang and Briggs, 1985). In some cases, a speedup above the ideal speedup can be obtained, conventionally known as super linear speedup, due to anomalies in programming, compilation, architecture usage, etc. For example, a single processor system may store all its data off-chip, whereas the multiprocessor system may store all its data on-chip providing an unpredicted increase in performance.

The efficiency (E_N) of a homogeneous parallel system is defined as

$$E_N = \frac{S_N}{N} \times 100\% = \frac{T_1}{NT_N} \times 100\% \tag{4.4}$$

Efficiency can be interpreted as providing an indication of the average utilisation of the 'N' processors, expressed as a percentage. Furthermore, this measure allows a uniform comparison of the various speedups obtained from systems containing different numbers of processors. It has also been illustrated that the value of efficiency is related to the granularity of the system (Stone, 1990). For example, consider an ideal problem which can be partitioned into N equal sub-tasks requiring R units of time, with associated communication overhead of C units. Thus, the ideal system's efficiency can be defined as

$$E_N = \frac{R/C}{1+(R/C)} \times 100\% = \frac{Granularity}{1+Granularity} \times 100\% \qquad (4.5)$$

Although, the analysis uses an ideal model, this value of granularity can be used as a guideline during the partitioning process.

Problem 4.1: *Find the speedup and efficiency of a homogeneous parallel architecture of five processors executing an algorithm. Assume that the execution time of the algorithm for a single processor in the architecture is 12 s, whereas, for the parallel architecture the execution time is 4 s.*

Solution: The speedup of the parallel architecture is

$$S_N = \frac{T_1}{T_N} = \frac{12}{4} = 3$$

where T_1 is the execution time for a single processor, N is the number of processors in the parallel architecture and T_N is the execution time for the N processors. Thus, the efficiency of the architecture is

$$E_N = \frac{S_N}{N} 100\% = \frac{3}{5} 100\% = 60\%$$

Problem 4.2: *Consider a homogeneous parallel architecture having an efficiency of 25% and speedup of 5. Find how many processors there are in the architecture. Determine the additional number of processors required so that an overall speedup of 8 is achieved.*

Solution: The efficiency of the architecture is

$$E_N = \frac{S_N}{N} 100\%$$

With efficiency of architecture $E_N = 25\%$ and speedup $S_N = 5$, the number of processors is obtained as

$$N = \frac{S_N}{E_N}100\% = \frac{5}{25\%}\times100\% = 5\times4 = 20$$

$$N = 20$$

Let P respresent the number of processors in the system achieving a speedup $S_P = 8$. Thus,

$$P = \frac{S_P}{E_P}100\% = \frac{8}{25\%}\times100\% = 8\times4 = 32$$

$$P = 32$$

Therefore, the additional number of processors required to achieve an overall speedup of 8 is

$$P - N = 32 - 20 = 12$$

Amdahl's Law

Gene Amdahl, an IBM designer derived a law for speedup of parallel processing in 1967. In brief, the law says that (Moldovan, 1993)

A small number of sequential operations can significantly limit the speedup achievable by a parallel computer

For example, if 5% of the operations of an algorithm must be performed sequentially, then the maximum speedup is 5, no matter how many processors a parallel computer has.

To explain this concept, assume that a parallel computer comprises N processors. Let η be the fraction of the algorithm that is to be executed sequentially. Thus, the parallel part of the algorithm will be $(1-\eta)$. As defined earlier in Equation 4.3, parallel speedup (S_N) is the ratio of the execution time (T_1) on a single processor, to the execution time (T_N) on N processors:

$$S_N = \frac{T_1}{T_N}$$

In this case, the serial part of the algorithm can be computed in a time equal to ηT_1 and the parallel part of the algorithm in a time $(1-\eta)T_1/N$. (In an ideal case, N

processors can execute the job in a fraction $1/N$ of the time of one processor.) Thus,

$$T_N = Execution\ time\ of\ sequentioal\ part$$
$$+ Execution\ time\ of\ parallel\ part\ of\ algorithm$$

or,

$$T_N = \eta T_1 + \frac{(1-\eta)T_1}{N} \tag{4.6}$$

Substituting this into the speedup Equation 4.3 yields

$$S_N \le \frac{1}{\eta + \dfrac{1-\eta}{N}} = \frac{N}{\eta(N-1)+1}$$

or,

$$S_N \le \frac{N}{\eta(N-1)+1} \quad (Amdahl's\ Law) \tag{4.7}$$

Note that as N approaches ∞ the speedup S_N approaches $1/\eta$.

Problem 4.3: *Consider an algorithm with a sequential part of 10% and parallel part of 90%. Determine the expected speedup that can be achieved in implementing this algorithm on a parallel architecture of (a) 10 processors, (b) 20 processors.*

Solution: Using Equation 4.7, the speedup can be evaluated for each case as

a. $S_N = \dfrac{10}{10\% \times (10-1)+1} = \dfrac{10}{\dfrac{10}{100} \times (10-1)+1} = 5.26$

b. $S_N = \dfrac{20}{10\% \times (20-1)+1} = \dfrac{20}{2.9} = 6$

4.4.2 Heterogeneous Architectures

Consider a heterogeneous parallel architecture of N processors. To define speedup and efficiency for the architecture, assume that a virtual processor is constructed with a performance, in terms of average speed, equivalent to the average performance of the N processors. Accordingly, let the performance characteristics of processor i ($i = 1,..., N$) over task increments of ΔW be given by

$$\Delta W = V_i \Delta T_i \tag{4.8}$$

where ΔT_i and V_i represent the execution time increment and average speed of the processor. Thus, the execution time increment ΔT_v and average speed V_v of the virtual processor executing the task increment ΔW can be obtained as

$$V_v = \frac{\Delta W}{\Delta T_v} = \frac{1}{N} \sum_{i=1}^{N} V_i \tag{4.9}$$

Substituting for V_i from Equation 4.8 into the above yields

$$\frac{\Delta W}{\Delta T_v} = \frac{1}{N} \sum_{i=1}^{N} \frac{\Delta W}{\Delta T_i} = \frac{\Delta W}{N} \sum_{i=1}^{N} \frac{1}{\Delta T_i}$$

Thus,

$$\Delta T_v = N \left(\sum_{i=1}^{N} \frac{1}{\Delta T_i} \right)^{-1} \tag{4.10}$$

Therefore, the fixed-load increment parallel speedup S_f and generalised parallel speedup S_g of the parallel architecture, over a task increment of ΔW, can be defined as

$$
\begin{aligned}
S_f &= \frac{\text{Execution time increment of virtual processor}}{\text{Execution time increment of parallel system}} = \frac{\Delta T_v}{\Delta T_p} \\
S_g &= \frac{\text{Average speed of parallel system}}{\text{Average speed of virtual processor}} = \frac{V_p}{V_v}
\end{aligned}
\tag{4.11}
$$

In this manner, the (fixed-load) efficiency E_f and generalised efficiency of the parallel architecture can be defined as

$$
\begin{aligned}
E_f &= \frac{S_f}{N} \times 100\% \\
E_g &= \frac{S_g}{N} \times 100\%
\end{aligned}
\tag{4.12}
$$

Note in the above that the concepts of parallel speedup and efficiency defined for heterogeneous architectures are consistent with the definitions given for homogeneous architectures. Thus, these can be referred to as the general definitions of speedup and efficiency of parallel architectures.

4.4.3 Task-to-Processor Allocation

The concept of generalised sequential speedup can be utilised as a guide to allocation of tasks to processors in parallel architectures so as to achieve maximum efficiency and maximum (parallel) speedup. Let the generalised sequential speedup of processor i (in a parallel architecture) to the virtual processor be $S_{i/v}$:

$$S_{i/v} = \frac{V_i}{V_v}; \quad i = 1,..., N \tag{4.13}$$

Using the processor characterisations of Equations 4.8 and 4.9 for processor i and the virtual processor, Equation 4.13 can alternatively be expressed in terms of fixed-load increment speedup as

$$S_{i/v} = \frac{\Delta T_v}{\Delta T_i}; \quad i = 1,..., N \tag{4.14}$$

Thus, to allow 100% utilisation of the processors in the architecture the task increments ΔW_i allocated to processors should be such that the execution time increment of the parallel architecture when implementing the task increment ΔW is given by

$$\Delta T_p = \Delta T_i = \frac{\Delta W_i}{V_i} = \frac{\Delta T_v}{N} = \frac{1}{N}\frac{\Delta W}{V_v}, \quad i = 1,..., N \tag{4.15}$$

or

$$\Delta W_i = \frac{V_i}{V_v}\frac{\Delta W}{N} = S_{i/v}\frac{\Delta W}{N}, \quad i = 1,..., N \tag{4.16}$$

It follows from Equation 4.15 that, with distribution of load among the processors according to Equation 4.16 the parallel architecture is characterised by

$$\Delta W = V_p \Delta T_p = (NV_v)\Delta T_p \tag{4.17}$$

having an average speed of

$$V_p = NV_v \tag{4.18}$$

Thus, with distribution of load among the processors according to Equation 4.16, the speedup and efficiency achieved with N processors are N and 100% respectively. These are the ideal speedup and efficiency. In practice, however, due to communication overheads and runtime conditions the speedup and efficiency of the parallel architecture will be less than these values.

Note in the above that, in developing performance metrics for a heterogeneous parallel architecture of N processors, the architecture is conceptually transformed into an equivalent homogeneous architecture incorporating N identical virtual processors. This is achieved by allocating tasks among the processors according to their computing capabilities to achieve maximum efficiency. For a homogeneous parallel architecture the virtual processor is equivalent to a single PE in the architecture.

4.5 Interpretations

The performance metrics developed above for parallel architectures are general in nature. In this section three possible scenarios relating to processor characterisation are considered and the corresponding procedures for performance evaluation of the architecture are outlined.

4.5.1 Linear Characterisation over Task Sizes Greater Than Zero

In the case of linear characterisation over $W \geq 0$, processor i ($i = 1,...,N$) is characterised by

$$W = V_i T_i, \quad i = 1,...,N \tag{4.19}$$

This, in relation to the characterisation in Equation 4.8, corresponds to having the reference task size and corresponding execution time at the point ($0,0$). Therefore, the general performance metrics and development of the parallel architecture can easily be formulated by taking out Δ throughout Equations 4.8 to 4.16. That is, interpreting increments as absolute quantities. In this manner, the corresponding virtual processor and (the most efficient) parallel architecture are characterised by

$$W = V_v T_v, \quad T_v = N \left(\sum_{i=1}^{N} \frac{1}{T_i} \right)^{-1} \tag{4.20}$$

and

$$W = V_p T_p = (NV_v) T_p \tag{4.21}$$

respectively. The parallel architecture in Equation 4.20 is achieved by the allocation of a task size W to processors according to

$$W_i = V_{i/v} \frac{W}{N}, \quad i = 1, \dots, N \tag{4.22}$$

The processor characterisations above all pass through the origin of the WT -plane.

Example 4.1
Consider three processors p_1, p_2 and p_3 characterised by

$$W = 1.5T_1, \quad W = 2T_2 \quad \text{and} \quad W = 4T_3$$

Thus, using Equations 4.9 and 4.20 the corresponding virtual processor characterising these processors in a parallel architecture is obtained with an average speed and characteristic of

$$V_v = \frac{V_1 + V_2 + V_3}{3} = \frac{1.5 + 2 + 4}{3} = 2.5, \quad W = 2.5T_v$$

The sequential speedups of processors p_1, p_2 and p_3 relative to the virtual processor are accordingly given as

$$S_{1/v} = 0.6, \quad S_{2/v} = 0.8 \quad \text{and} \quad S_{3/v} = 1.6$$

Thus, using Equation 4.20 the architecture can, conceptually, be transformed into an equivalent homogeneous architecture of three virtual processors by allocating the tasks among processors p_1, p_2 and p_3 as

$$W_1 = \frac{W}{5}, \quad W_2 = \frac{4W}{15} \quad \text{and} \quad W_3 = \frac{8W}{15}$$

In this manner, using Equation 4.21 the execution time of the parallel architecture in implementing a task size of W is given by

$$T_p = \frac{2W}{15} = \frac{T_v}{3}$$

Figure 4.1 shows the characteristics of the above processors. Thus, the speedup achieved with the parallel architecture relative to processors p_1, p_2, p_3 and the virtual processor are 5, 4.75, 1.875 and 3 respectively.

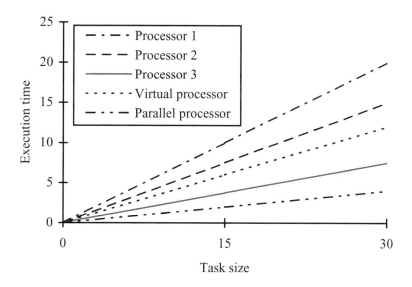

Figure 4.1. Characteristics of the processors

4.5.2 Linear Characterisation over a Range of Task Sizes

In the case of linear characterisation over $W \geq W_0$, processor i ($i = 1,...,N$) is characterised by Equation 4.8. This can be expressed alternatively as

$$W = V_i T_i + b_i, \quad i = 1,...,N \quad \text{for } W \geq W_0 \tag{4.23}$$

The corresponding virtual processor is characterised by

$$W = V_v T_v + b_v \quad \text{for } W \geq W_0 \tag{4.24}$$

In this case, if W approaches zero ($W \rightarrow 0$) then, according to Equations 4.10 and 4.22, it is expected that

$$T_i \rightarrow -\frac{b_i}{V_i}, \quad T_v \rightarrow -N\left(\sum_{i=1}^{N}\frac{V_i}{b_i}\right)^{-1} \quad \text{for } W \rightarrow 0 \tag{4.25}$$

Thus, it can be deduced from Equations 4.23 and 4.24 that

$$b_v = -NV_v \left(\sum_{i=1}^{N} \frac{V_i}{b_i} \right)^{-1} \tag{4.26}$$

In this manner, the corresponding (most efficient) parallel architecture can be constructed using the task allocation in Equation (4.16), resulting in the characterisation

$$W = V_p T_p + b_p \quad \text{for } W \geq W_0 \tag{4.27}$$

where,

$$V_p = NV_v \quad \text{and} \quad b_p = b_v$$

Note that, since the virtual processor and the parallel processor characterisations both pass through the same point for $T = 0$ in the WT-plane, the speedup and efficiency of the parallel architecture can be evaluated by using either fixed tasks or task increments.

Example 4.2
Consider three processors p_1, p_2 and p_3 characterised by

$$W = 2T_1 - 0.4, \quad W = 4T_2 - 2 \quad \text{and} \quad W = 6T_3 - 6 \quad \text{for } W \geq 3$$

Thus, using Equations 4.9 and 4.25 the corresponding virtual processor characterising these processors in a parallel architecture is obtained with an average speed and characteristic of

$$V_v = \frac{V_1 + V_2 + V_3}{3} = \frac{2 + 4 + 6}{3} = 4, \quad W = 4T_v - 1.5 \quad \text{for } W \geq 3$$

The sequential speedups of processors p_1, p_2 and p_3 relative to the virtual processor are accordingly given as

$$S_{1/v} = 0.5, \quad S_{2/v} = 1 \quad \text{and} \quad S_{3/v} = 1.5$$

Thus, using Equation 4.16 the architecture can, conceptually be transformed into an equivalent homogeneous architecture of three virtual processors by allocating the task increments among processors p_1, p_2 and p_3 as

$$\Delta W_1 = \frac{\Delta W}{6}, \quad \Delta W_2 = \frac{\Delta W}{3} \quad \text{and} \quad \Delta W_3 = \frac{\Delta W}{2}$$

In this manner, using Equation 4.17 the execution time increment of the parallel architecture when implementing a task increment of ΔW is given by

$$\Delta T_p = \frac{\Delta W}{12} = \frac{\Delta T_v}{3}$$

Thus, the speedup achieved with the parallel architecture relative to the processors p_1, p_2, p_3 and the virtual processor are 6, 3, 2 and 3 respectively. The architecture thus obtained is characterised by

$$W = 12T_p - 1.5 \quad \text{for } W \ge 3$$

Figure 4.2 shows the characteristics of the above processors.

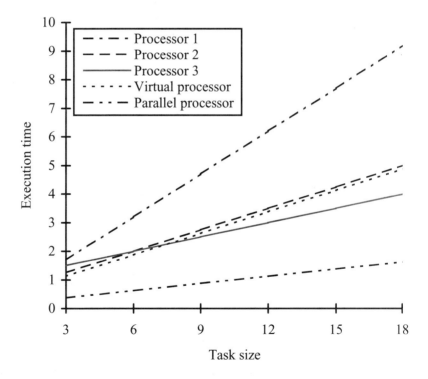

Figure 4.2. Characteristics of the processors

4.5.3 Piece-wise Linear Characterisation

In this case, processor i ($i = 1,...,N$) is characterised as Equation 4.8 over certain ranges of task sizes. This can be expressed, over a certain range of interest (W_1, W_2), as

$$W = V_i T_i + b_i, \quad i = 1,...,N \quad \text{for } W_1 \le W \le W_2 \tag{4.28}$$

The corresponding virtual processor and parallel architecture can be constructed using the procedure outlined in section 4.5.2 as

$$\begin{aligned} W &= V_v T_v + b_v \\ W &= V_p T_p + b_p \end{aligned} \quad \text{for } W_1 \le W \le W_2 \tag{4.29}$$

where,

$$\begin{aligned} V_v &= \frac{1}{N} \sum_{i=1}^{N} V_i, \quad b_v = -NV_v \left(\sum_{i=1}^{N} \frac{V_i}{b_i} \right)^{-1} \quad \text{for } W_1 \le W \le W_2 \\ V_p &= NV_v, \qquad b_p = b_v \end{aligned}$$

In this manner, the virtual processor and parallel architecture characterisations can be constructed over other ranges of sizes accordingly so as to cover the full range of interest.

Example 4.3
Consider three processors p_1, p_2 and p_3 characterised by

$$W = \begin{cases} 4T_1 & \text{for } 0 \le W < 5 \\ T_1 + 3.75 & \text{for } 5 \le W \end{cases}$$

$$W = \begin{cases} 8T_2 & \text{for } 0 \le W < 10 \\ 5T_2 + 3.75 & \text{for } 10 \le W \end{cases}$$

$$W = \begin{cases} 3T_3 & \text{for } 0 \le W < 15 \\ 1.2T_3 + 7.5 & \text{for } 15 \le W \end{cases}$$

Thus to construct the corresponding virtual processor and allocate tasks to processors accordingly, it is required to consider the above characterisations over the ranges $0 \le W < 5$, $5 \le W < 10$, $10 \le W < 15$, $0 \le W < 5$ and $5 \le W$.

For $0 \le W < 5$:

The processors p_1, p_2 and p_3 are characterised over this range by

$$W = 4T_1, \quad W = 8T_2 \quad \text{and} \quad W = 3T_3 \quad \text{for } 0 \le W < 5$$

Thus, using Equations 4.9 and 4.10, the speeds and characterisations of the corresponding virtual processor and the parallel architecture are obtained over this range as

$$V_v = \frac{4+8+3}{3} = 5, \quad W = 5T_v \quad \text{for } 0 \le W < 5$$

$$V_p = 3V_v = 15, \quad W = 15T_p \quad \text{for } 0 \le W < 5$$

At $W = 5$ the above equations yield $T_1 = 1.25$, $T_2 = 0.625$, $T_3 = 5/3$, $T_v = 1$ and $T_p = 1/3$. The speedups achieved with the parallel architecture relative to processors p_1, p_2, p_3 and the virtual processor over this range of task sizes are 4.75, 1.87, 5 and 3 respectively.

For $5 \le W < 10$:

The processors p_1, p_2 and p_3 are characterised over this range by

$$W = T_1 + 3.75, \quad W = 8T_2 \quad \text{and} \quad W = 3T_3 \quad \text{for } 5 \le W < 10$$

Using Equations 4.9 and 4.29 the speeds and characterisation of corresponding virtual processor and the parallel architecture are obtained over this range as

$$V_v = \frac{1+8+3}{3} = 4, \quad W = 4T_v + b_v \quad \text{for } 5 \le W < 10$$

$$V_p = 3V_v = 12, \quad W = 12T_p + b_p \quad \text{for } 5 \le W < 10$$

where, using $T_v = 1$ and $T_p = 1/3$ at $W = 5$ yields $b_v = b_p = 1$. Thus

$$\begin{aligned} W &= 4T_v + 1 \\ W &= 12T_p + 1 \end{aligned} \quad \text{for } 5 \le W < 10$$

At $W = 10$ the above equations yield $T_1 = 6.25$, $T_2 = 1.25$, $T_3 = 10/3$, $T_v = 2.25$ and $T_p = 0.75$. Thus, the speedups achieved with the parallel architecture relative to processors p_1, p_2, p_3 and the virtual processor over this range of task sizes are 12, 1.5, 4 and 3 respectively.

For $10 \leq W < 15$:

The processors p_1, p_2 and p_3 are characterised over this range by

$$W = T_1 + 3.75, \quad W = 5T_2 + 3.75 \quad \text{and} \quad W = 3T_3 \quad \text{for } 10 \leq W < 15$$

Using Equations 4.9 and 4.29 the speeds and characterisations of the corresponding virtual processor and the parallel architecture are obtained over this range as

$$V_v = \frac{1+5+3}{3} = 3, \quad W = 3T_v + b_v \quad \text{for } 10 \leq W < 15$$

$$V_p = 3V_v = 9, \quad W = 9T_p + b_p \quad \text{for } 10 \leq W < 15$$

where, using $T_v = 2.25$ and $T_p = 0.75$ at $W = 10$ yields $b_v = b_p = 3.25$. Thus

$$\begin{aligned} W &= 3T_v + 3.25 \\ W &= 9T_p + 3.25 \end{aligned} \quad \text{for } 10 \leq W < 15$$

At $W = 15$ the above equations yield $T_1 = 11.25$, $T_2 = 2.25$, $T_3 = 5$, $T_v = 11.75/3$ and $T_p = 11.75/9$. The speedups achieved with the parallel architecture relative to processors p_1, p_2, p_3 and the virtual processor over this range of task sizes are 9, 1.8, 3 and 3 respectively.

For $15 \leq W$:

The processors p_1, p_2 and p_3 are characterised over this range by

$$W = T_1 + 3.75, \quad W = 5T_2 + 3.75 \quad \text{and} \quad W = 1.2T_3 \quad \text{for } 15 \leq W$$

Using Equations 4.9 and 4.29 the speeds and characterisations of the corresponding virtual processor and the parallel architecture are obtained over this range as

$$V_v = \frac{1+5+1.2}{3} = 2.4, \quad W = 2.4T_v + b_v \quad \text{for } 15 \leq W$$

$$V_p = 3V_v = 7.2, \quad W = 7.2T_p + b_p \quad \text{for } 15 \leq W$$

where, using $T_v = 11.75/3$ and $T_p = 11.75/9$ at $W = 15$ yields $b_v = b_p = 5.6$. Thus

$$W = 2.4T_v + 5.6$$
$$W = 7.2T_p + 5.6$$
for $15 \leq W$

Thus, the speedups achieved with the parallel architecture relative to processors p_1, p_2, p_3 and the virtual processor over this range of task sizes are 7.2, 1.44, 6 and 3 respectively. Figure 4.3 shows the piece-wise linear characterisations of the above processors. Note that for each linear range it is assumed that each PE of the parallel architecture behaves according to its characterisation for that range of task sizes. As mentioned earlier, the anomalies in practice due to a change of speed within the characterisation of a processor is not significant; this assumption does not lead to substantial errors in the theoretical characterisation of the virtual processor and the parallel processor.

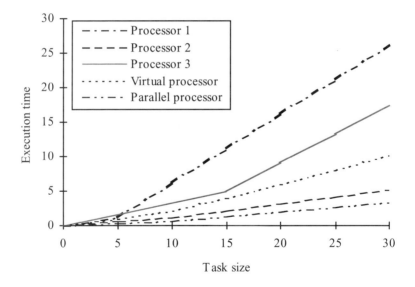

Figure 4.3. Characteristics of the processors

Problem 4.4: *Consider an algorithm with the task dependencies given in Figure 4.4. This is implemented on a homogeneous parallel architecture of three processors P_i (i=1,2,3). Let the corresponding execution times of the tasks on a single processor be as in Table 4.1. Allocate the tasks to the processors for maximum efficiency and determine the* *resulting idle time of each processor, and speedup and efficiency of the parallel architecture. Assume zero communication overhead.*

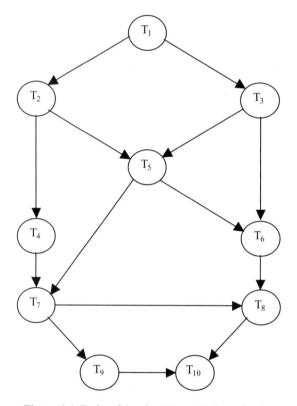

Figure 4.4. Tasks of the algorithm with dependencies

Solution: The tasks of the algorithm can be allocated to the processors P1, P2 and P3 as shown in Table 4.2: then the total execution time of the architecture is 25 s, whereas the time required for a single processor is 39 s. It is noted from Table 4.2 that the total idle time of the three processors is 36 sec and, on the other hand, the utilisation time of the processors is 39 sec. Thus, the speedup of the parallel architecture with three identical processors is

$$S_N = \frac{T_1}{T_N} = \frac{39}{25} = 1.56$$

where T_1 is the execution time for a single processor, N is the number of processors in the parallel architecture and T_N is the execution time for N processors. Thus, the efficiency of the architecture is

$$E_N = \frac{S_N}{N}100\% = \frac{1.56}{3}100\% = 52\%$$

Table 4.1. Performance of processors for different algorithm tasks

Task	Execution time (s) of single processor (P_i type)
T_1	4
T_2	3
T_3	5
T_4	4
T_5	4
T_6	3
T_7	4
T_8	5
T_9	4
T_{10}	3

4.6 Case Study

To demonstrate the application of the performance metrics described earlier in a practical context, a series of experiments have been conducted and the results are presented and discussed in this section. The algorithms utilised include a finite difference beam simulation and an add and multiply algorithm referred to as the DOT algorithm. Full details of these algorithms are given in Chapter 6, and the hardware and software resources utilised are described in Appendix B. In these case studies, with the beam simulation algorithm, an aluminium type cantilever beam was considered. The beam was divided into 19 segments and a sample period of $\Delta t = 0.3$ ms was used. The total execution times achieved by the architectures when implementing the simulation algorithm over 20,000 iterations were considered.

Table 4.2. Task allocation table of a parallel architecture with three identical P_i processors

Time (s)	P1	P2	P3
1	T_1		
2			
3			
4			
5		T_2	T_3
6			
7			
8		T_4	
9			
10	T_5		
11			
12			
13			
14		T_7	T_6
15			
16			
17			
18		T_9	
19			
20			T_8
21			
22			
23	T_{10}		
24			
25			

4.6.1 Sequential Computing

The flexible beam simulation algorithm was implemented as a sequential process on uniprocessor-based architectures. The total execution times achieved by the architectures when implementing the simulation algorithm over 20,000 iterations were measured. It was noted that among the processors used the Pentium PII (300 MHz) processor performed the fastest and the transputer T8 processor the slowest of the processors used. Table 4.3 shows the performance of the computing platforms used relative to that of the Pentium PII. The simulation algorithm, as discussed later in Chapter 6, is mainly of a matrix-based computational type for which the powerful vector processing resources of the i860 are exploited and utilised to achieve the shortest execution time among the processors other than the PII. The C40 does not have such vector processing resources making it about 87 times and 6 times slower than the PII and the i860, respectively. This implies that the C40 is not performing well in a situation where the algorithm is of matrix type and extensive runtime memory management is involved. A single transputer, on the other hand, performed about 146 times slower than the PII. The SPARC and the 486DX2 appear to have achieved similar performances, with the SPARC being slightly faster due to its RISC processor. The T8 performed the slowest of all the processors. This is mainly due to floating-point operations, which are evaluated in software rather than using dedicated hardware since it has no maths co-processor. Moreover, it does not have internal cache memory (although it has a very small internal RAM) making it slower when handling calculations of large amounts of data.

Table 4.3. Execution times of various computing platforms (T1) relative to the vector processor Pentium PII (300 MHz) (T2) in implementing the simulation algorithm

	i860	C40	T8	SPARC	Pentium 90 MHz	486DX2
T1/T2	14.0	86.0	146.0	41.0	14.5	56.0

4.6.2 Homogeneous Parallel Computing

To investigate real-time implementation of the simulation algorithm, the algorithm was implemented on a transputer network of up to nine T8s, on the basis of the case study provided in Section 3.4.1.

The speedup and corresponding efficiency of the execution time for networks of up to nine T8s are shown in Table 4.4. As discussed earlier, an increase in the number of transputers results in a decrease in the execution time. The relation, however, is not always linear. This implies that, with an increase in the number of processors or a transformation of the algorithm from course-grain to fine-grain, more and more communication is demanded. This is evidenced in Table 4.4, which shows the non-linear variation in speedup and efficiency. This implies that the algorithm considered as a fixed load is not suitable for exploitation with this hardware, due to communication overheads and runtime memory management problems.

Figure 4.5 and Table 4.5 show the execution times and corresponding speedups and efficiencies of the network of C40s when implementing the algorithm. It is noted that the execution time, speedup and efficiency achieved with the network of C40s is not linear. Compared to the network of T8s, the network of C40s achieves better performance. However, due to a mismatch between the algorithm and architecture, communication overhead and runtime memory management problem, the network of C40s has not achieved outstanding speedups and efficiencies.

Table 4.4. Speedup and efficiency of the T8 network for the simulation algorithm

Number of T8s	Two	Three	Four	Five	Six	Seven	Eight	Nine
Speedup	1.55	1.73	1.935	1.943	2.1	2.15	2.174	2.19
Efficiency	77%	58%	48%	39%	35%	31%	27%	24%

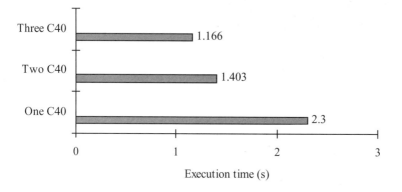

Figure 4.5. Execution time of the network of C40s when implementing the beam simulation algorithm

Table 4.5. Speedup and efficiency of the C40s network when implementing the beam simulation algorithm

Number of C40s	Two	Three
Speedup	1.64	1.973
Efficiency	82.00%	65.77%

4.6.3 Heterogeneous Parallel Computing

In this section results of investigations to verify the task allocation strategy are presented through implementing the beam simulation and DOT algorithms on heterogeneous architectures. In implementing the beam simulation algorithm, algorithm granularity was achieved by increasing the number of beam segments from 5 to 40, in steps of 5. The execution times of the processors were obtained over 20000 iterations at each step. In the case of the DOT algorithm the number of data points was varied from 1000 to 10,000, in steps of 1000. For simplicity, the vectors $b(i)$ and $c(i)$ were fixed at 4.0 and 6.5 respectively.

To implement the DOT algorithm, the heterogeneous architecture of the T8 and C40 was utilised. Task to processor allocation was achieved according to the strategy described in Section 4.4.3. Figure 4.6 shows the execution times achieved with the C40, T8 and C40+T8 architectures when implementing the DOT algorithm. The characteristics of the virtual processor and the corresponding theoretical C40+T8 heterogeneous architecture are also shown. It is noted that the performance of the actual C40+T8 is close to its corresponding theoretical (100% efficient) model. As there is no interprocessor communication involved in implementing the algorithm, the generalised speedup and efficiency achieved with the C40+T8 are nearly 2 and 100% respectively.

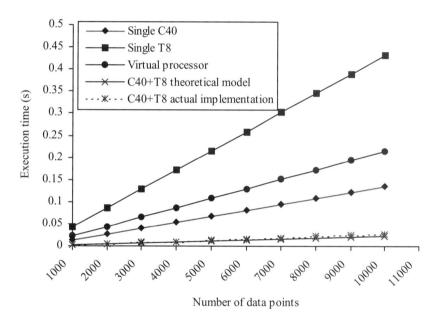

Figure 4.6. Execution times of the C40+T8 architecture when implementing the DOT algorithm

Figure 4.7 shows the execution times of the C40, T8 and C40+T8 architectures when implementing the beam simulation algorithm. The characteristics of the virtual processor and the corresponding theoretical C40+T8 heterogeneous architecture are also shown in Figure 4.7. It is noted that among the uni-processors, the C40 is considerably faster than the T8. The combination thus results in a virtual processor with characteristics closer to those of the C40. This implies that, for the C40+T8 to achieve maximum efficiency, a large proportion of the task must be allocated to the C40. It is noted that the actual C40+T8 performed slower than the corresponding theoretical (100% efficient) model. This is due to the communication between processors.

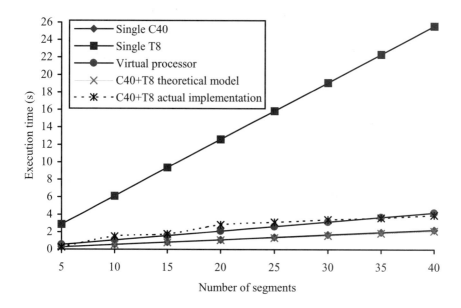

Figure 4.7. Execution times of the C40+T8 architecture when implementing the beam simulation algorithm

Figure 4.8 shows the execution times of the i860, T8 and the i860+T8 architectures when implementing the beam simulation algorithm. The characteristics of the virtual processor and the corresponding theoretical i860+T8 heterogeneous architecture are also shown. It is noted that the i860 is much faster than the T8, and for this reason all the task was allocated to the i860, with the exception of the last two values (35 and 40 segments) where one segment was allocated to the T8. The performance of the actual i860+T8 was similar to the uni-processor experiment with the i860, except for the last two values where interprocessor communication increased the execution time for the actual implementation.

Figure 4.9 shows the execution times of the C40, i860 and i860+C40 heterogeneous architectures when implementing the beam simulation algorithm. The characteristics of the virtual processor and the corresponding theoretical i860+C40 heterogeneous architecture are also shown. It is noted that the actual implementation on the i860+C40 is faster than the virtual processor and faster than the i860. It is also noted that the actual implementation is very close to the virtual (theoretical) parallel machine.

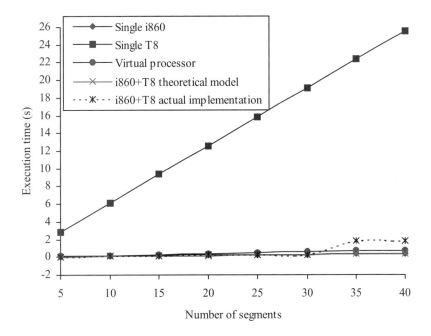

Figure 4.8. Execution times of the i860+T8 architecture when implementing the beam simulation algorithm

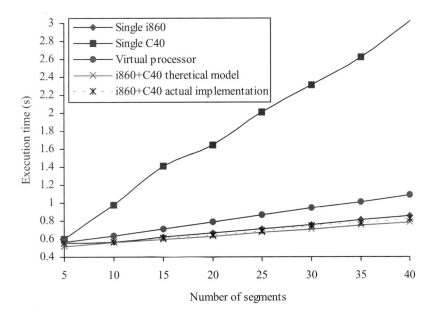

Figure 4.9. Execution times of the i860+C40 architecture when implementing the beam simulation algorithm

Comparing the three implementations of the beam simulation algorithm, it is noted that with the C40+T8, the execution times of the actual implementation stay very close to the virtual processor. Task allocation was computed according to the theoretical model. However, communication overheads are evident with the i860+T8 implementation, where no task was allocated to the T8 except for the last two segment numbers. This resulted in the same execution times as the i860 processor, except for the 35 and 40 segment steps, where the execution time increased. The results of these two implementations show that due to a disparity in the capabilities of the processors, communication overhead becomes a dominant factor in the implementation. The i860+C40 implementation shows that a better task allocation is obtained between both processors, and communication overheads are minimum. In this manner the best performance for this algorithm was obtained with the combination i860+C40.

4.7 Summary

The performance metrics for sequential and parallel computing domains have been introduced and presented with a number of case studies. A generalised speedup as a quantitative measure of performance evaluation of processors has been introduced. A task allocation strategy allowing maximum efficiency has been developed using the concept of generalised speedup. Application of the concepts has been demonstrated through theoritical examples and practical case studies.

4.8 Exercises

1. Distinguish between speedup in sequential and parallel computing? Indicate condition(s) under which superlinear speedup may be achieved with homogeneous parallel processing?

2. Briefly describe the benchmark of an architecture? Distinguish between efficiency and isoefficiency. Indicate what is meant by speedup in sequential processing?

3. Define the speedup and efficiency of homogeneous and heterogeneous architectures in a parallel processing context. Consider the efficiency of a homogeneous parallel architecture to be 50% and its speedup 4; determine the number of processors in the architecture. Determine the number of processors required for a speedup of 8 and an efficiency of 50% to be achieved with a homogeneous parallel architecture.

4. Allocate the tasks shown in Figure 4.10 to processors of a homogeneous parallel architecture comprising four identical processors P_i ($i = 1,2,3,4$). Let the performance of each processor executing the tasks be as given in

Table 4.6. Find the idle time of the processors, and the speedup and efficiency of the architecture for the algorithm with zero communication overhead.

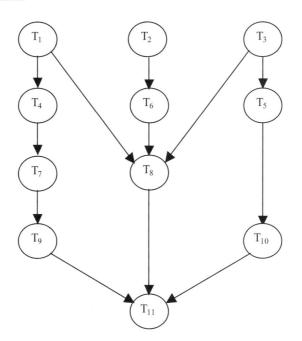

Figure 4.10. Algorithm tasks with dependencies

Table 4.6. Processor performance for different tasks of an algorithm

Task	Execution time (s) of single processor (type P_i)	Execution time (s) of single processor (type P_j)
T_1	4	2
T_2	3	1.5
T_3	5	2.5
T_4	4	2
T_5	4	2
T_6	3	1.5
T_7	4	2
T_8	5	2.5
T_9	4	2
T_{10}	3	1.5
T_{11}	4	2
T_{12}	4	2

5. Allocate the tasks shown in Figure 4.10 to processors of a heterogeneous parallel architecture of two P_i processors and one P_j processor to achieve maximum performance of the architecture. The performances of the processors reflecting different tasks are given in Table 4.6. Find the idle time of the processors, and the speedup and efficiency of the architecture for the algorithm with zero communication overhead.

5. Parallel Programming and Multithreading

5.1 Objectives

- To introduce the concept of parallel programming.
- To understand parallel programming languages.
- To introduce parallel programming models.
- To explore multithreading for a multiprocessor computing domain.
- To demonstrate real-time implementation issues using multithreading concepts.

5.2 Introduction

This chapter presents software and, in turn, programming aspects of parallel processing for real-time computing. This includes introduction to the basic terminology for parallel programming, parallel programming languages, parallel programming models and concurrent multithreading. Concurrent multithreading is currently a growing technology for generic parallel and distributed computing. This chapter covers real-time issues for multithreading in a multiprocessor domain with emphasis on thread synchronisation.

A program is a sequence of instructions that direct a computing domain to perform certain functions. A function of a program is a collection of declarations and statements that carries out a specific action and/or returns a value. Functions are either defined by the user, or have been previously defined and made available to the user. An instruction consists of one or more statements, where a statement might contain one or more arithmetic or logical operations with constants and/or variables.

A process is part of a program, defined as a sequence of program instructions that can be performed in sequence or in parallel with other processes of the program or program instructions. A program can be divided into a number of processes, which are executed sequentially or concurrently. In traditional sequential programming all the processes execute sequentially although there might be instruction-level parallelism depending on the compiler, operating system and the PE. However, users do not have any control over this.

In multiprogramming, the processing unit has control transferred from one program to another, causing their instructions to be interleaved at stages of their execution. In the case of multiprogramming in a multiprocessor environment, more than one program can be executed at the same time, meaning that each program proceeds independently with its execution process. On the other hand, within a parallel processing environment, processes of a single program can be executed in parallel in a number of processors. In such a system, processes might interact and affect each other, based on the data dependencies and control dependencies of the processes. Therefore, users need to provide additional effort to execute a parallel program in a parallel architecture compared to a sequential program in a single PE. This effort includes process mapping and scheduling, interprocess synchronisation and interprocessor network configuration. Although a few operating systems and software can handle some of these issues dynamically to a certain extent, efficient dynamic scheduling and synchronisation is still under investigation for further levels of development.

Selecting a suitable programming language for parallel computing is an important and challenging issue. Some key properties of parallel programming languages are compatibility, expressiveness and ease of use, efficiency and portability. In order to exploit the parallelism, programmers need to look at all these issues for a suitable solution. A programming language that contains explicit mechanisms for parallel processing must be constructed to create a new process.

The concept of concurrent multithreading in uniprocessor and multiprocessor computing is currently a subject of widespread interest among scientists and professional software developers. For some specific applications, concurrent multithreading offers better performance even in a single-processor-based architecture, particularly where calculations and I/O operations run in parallel without any dependency. However, the key interest in this chapter is to concentrate on concurrent multithreading for real-time implementation of signal processing and control algorithms on multiprocessor domains where data and control dependencies are relatively high.

Concurrent multithreading in a multiprocessor domain is a growing technology, with the potential for enabling the development of much higher performance applications compared to the traditional method of software development in uniprocessor or multiprocessor environments. In the mid-to-late 1980s, several research projects focused on threaded programming designs. In the early 1990s, threads were established in various UNIX operating systems, such as the USL's system V Release 4, Sun Solaris, and the Open Software Foundation's OSF/1. However, the platform-specific threads programming libraries highlighted the need for some portable, platform-independent threads interface. Recently, the IEEE has recognised this need with the acceptance of the IEEE standard for portable operating system interface (POSI). However, a considerable amount of research work has been devoted recently to the development methodologies of concurrent multithreading (Asche, 1996; Gray, 1998; Lo et al., 1998; Nichols et al., 1996; Redstone et al., 2000; Tullsen et al., 1999).

5.3 Parallel Programming Languages

Parallel programming languages are mostly extensions of sequential programming languages that are used in single processor based architectures. In general, these programming languages have provided a particular model of parallel organisation in providing parallel programming facilities. Some languages use the shared-memory model and provide facilities for mutual exclusion, whereas others assume the distributed model and provide link-based communication facilities. A few languages have included both models, because different models are preferable in particular solutions. However, parallel programming languages still lack some essential entities for parallel processors such as synchronisation, communication protocols, granularity, network control and masking.

In general, a programming language must be machine independent and must adhere to language design principles such as readability, writeability, maintainability, and portability. Moreover, efficiency of a particular programming language depends on the efficiency of compilers, code optimisation capability and dynamic memory management facilities (Hossain, 1995). However, like sequential programming languages, parallel programming languages can also be classified as procedural programming languages and non-procedural programming languages (Braunl, 1993). These are briefly discussed below.

5.3.1 Procedural Programming Languages

Procedural programming languages include asynchronous and synchronous procedural programming languages. In an asynchronous form of parallelism, the program for a problem to be solved contains multiple threads of control. In this case, if the processes (subprograms) are not totally independent, the processes must exchange information among them and therefore must be mutually synchronised. This is mostly suitable for large-grained or program-level parallelism in MIMD architecture, particularly where a huge overhead might be incurred if the program is implemented in a fine-grain form. Some of the most common asynchronous procedural parallel programming languages for the MIMD computing domain are: Concurrent Pascal, Occam, Ada, Sequent-C and Linda.

A synchronous parallel program, particularly for an SIMD architecture, contains only a single thread of control, where a central PE controls all the processors with a global clock. In this case, parallel process synchronisation is relatively easy and fine grain is suitable, although the granularity of the program depends on data and control dependencies. Some of the most popular synchronous procedural parallel programming languages for SIMD computing are Fortran 90, Fortran D, C-Star, Parallel C, MasPar and Parallaxis.

5.3.2 Non-procedural Programming Languages

Non-procedural parallel programming languages are also known as functional and logic programming languages. Non-procedural programming languages are gaining importance day by day. In general, non-procedural programming languages exhibit a higher level of abstraction than procedural languages. However, with non-procedural programming languages, the handling of parallelism can be completely transparent to the user. Some of the most common non-procedural parallel programming languages are Star-Lisp, FP (functional program) and Concurrent Prolog.

5.4 Parallel Programming Model

Programming models for sequential computing have a large number of languages and algorithms based on the von Neumann architecture. With the evolution of parallel hardware, parallel programming languages have come to support the specification and management of parallelism. Initially, these did not receive the acceptance that most sequential programming languages had and they remain difficult to use due to several shortcomings including lack of machine independence, scalability, and programming support. In general, the primary challenge of hardware and software designers alike is to provide an environment where parallel programs can be designed, implemented, and debugged by programmers who have both natural and learned tendencies to decompose problems in a sequential manner. The most common approaches to meet this challenge are sequential programming, asynchronous parallel programming and synchronous parallel programming models.

A sequential programming model presents a traditional sequential program and has an automated system, which converts it to parallel code. A program is first analysed to determine a partial ordering of its operations on data that will preserve the correctness (meaning) of the sequential program. Although this approach may provide the eventual solution, it requires very complex program analysis.

The asynchronous parallel programming model represents a program as a set of sequential processes that occasionally interact with each other. In a program, processes are written based on the assumption that all processes execute asynchronously. If one process needs information from another, it may be forced to wait for that information to become available, but no other processes are affected by this interaction. This model is generally used in conjunction with MIMD execution models.

The synchronous parallel programming model allows parallelism only within a given step of a solution. A program specifies a sequence of such steps, with parallelism facilitated by simultaneous execution of each step in many processors. This approach is often associated with an SIMD execution model. In such a model, only a single sequence of instructions is present, but each instruction may be simultaneously executed in an arbitrary set of processors, with each processor operating upon its own data.

It is worth noting that the approaches above map the problem of programming for parallel execution into a sequential domain. Synchronous models allow the problem to be decomposed into a series of small steps that may be executed in parallel, while asynchronous models treat the problem as a set of nearly independent sequential tasks that can proceed concurrently. Although the sequential technique can potentially be applied in conjunction with the other two approaches, synchronous and asynchronous models have typically been considered to be mutually exclusive.

5.5 Multithreading in Multiprocessing

Threads reduce overheads by sharing fundamental parts, for instance, code (text), data, stack and file I/O. By sharing these parts switching happens much more frequently and efficiently. Although sharing information is not difficult, sharing with a dependency between threads causes degradation of the overall computing performance. A single-thread program possesses a single execution thread or library not designed to be used in a multithread environment. A multithread program uses multiple execution threads or a library that allows concurrent access from multiple threads. Concurrent multithreading separates a process into many execution threads, each of which runs independently. Typically, applications that express concurrency requirements with threads need not take into account the number of available processors. The performance of the application improves transparency with additional processors. Numerical algorithms and applications with a high degree of parallelism, such as matrix multiplication, can run faster when implemented with threads on a multiprocessor (Gray, 1998).

Dependency is one of the key issues in multithreading for real-time high-performance computing. During the implementation of an algorithm, data dependency between two blocks or two statements requires memory access time. In practice, increasing dependencies imply increasing access time or more inter-process communication in the case of concurrent multithread implementation in a multiprocessor domain. This, accordingly, degrades the real-time performance. Thus, it is essential to study and analyse data dependencies in an algorithm intended for real-time concurrent thread implementation. Such a study must address essential questions, such as how to reduce the block or statement dependencies, and what is the impact of data dependencies in real-time inter-process communication? Detection of multithreading in an application involves finding sets of computations that can be performed simultaneously. The approach to parallel multithreading is thus based on the study of data dependencies. The presence of dependence between two computations implies that they cannot be performed in parallel. In general, the fewer the dependencies, the greater the parallelism (Moldovan, 1993). Many algorithms have regular data dependencies, that is, dependencies that repeat over the entire set of computations in the algorithm. For such algorithms, dependencies can be concisely described mathematically and can be manipulated easily. However, there are algorithms for which dependencies vary from one computation to another and these algorithms are more difficult to analyse. When two or more algorithms have similar dependencies, it means that they exhibit similar parallel properties.

Synchronisation is another key issue of concurrent multithreading in a multiprocessor domain for real-time computing. The performance of the synchronisation mechanism of a multiprocessor determines the granularity of parallelism that can be exploited on that machine. Synchronisation on a multiprocessor carries a high cost due to the hardware levels at which synchronisation and communication must occur (Tullsen *et al.*, 1999). Therefore, the study of concurrent multithreading includes process synchronisation, inter-process communication, scheduling and mapping of processes and granularity. These are discussed below.

5.5.1 Thread or Process Synchronisation

Parallel multithreading offers one way to solve complex computational problems quickly by creating and co-ordinating multiple execution processes. In practice, the execution of a parallel program forms a set of concurrent execution process (threads), which communicate and synchronise by reading and writing shared variables. There must be a way to coordinate the activities of these processes. For instance, consider a program that contains three routines. Two routines write to variables and the third reads them. For the final routine to read the right values, one must add some synchronisation. It is worth noting that by using finer synchronisation techniques, threads can spend less time waiting for each other and more time accomplishing the tasks for which they were designed.

In multithreading, sometimes the processes work on their own data and do not interact. But processes must communicate and synchronise with each other when they exchange results of the execution. In general, there are two methods of synchronisation namely, synchronisation for precedence and synchronisation for mutual exclusion that can be employed.

The method of synchronisation for precedence guarantees that one event does not begin until another event has finished. The method of synchronisation for mutual exclusion, on the other hand, guarantees that only one process can access the critical section where the data are shared and must be manipulated. To implement these methods, one has to use the mechanism called semaphores and/or mutual exclusion.

Semaphores can be briefly defined as a locking mechanism, basically used for file locking. One can use them to control access to files, shared memory and I/O devices. The basic functionality of a semaphore is that one can either set it, check it, or wait until it clears then set it.

In concurrent multithreading, processes that are working together often share some common storage that one can read and write to. The shared storage may be in the main memory or it may be a shared file. Each process has a segment of code, called a critical section, which accesses shared memory or files. The key issue involving shared memory or shared files is to find a way to prohibit more than one process from reading and writing the shared data at the same time. This requires mutual exclusion, as a way of making sure that if one process is executing in its critical section, other processes will be excluded from doing the same thing.

Classic synchronisation mechanisms are mutexes/condition variables and semaphores. These are provided in the kernel of an operating system, together with other synchronisation/communication mechanisms that are common in real-time systems, such as event flags and message queues.

In real-time systems priority inversion is one of the key problems one has to deal with. This is where a high priority thread is (wrongly) prevented from continuing by one with lower priority. A normal example is of a high-priority thread waiting at a mutex already held by a low-priority thread. If the low-priority thread is pre-empted by a medium-priority thread then priority inversion has occurred since the high-priority thread is prevented from continuing by an unrelated thread of lower priority. Within a number of solutions, the simplest is to employ a priority ceiling protocol where all threads that acquire the mutex have their priority boosted to some predetermined value. This has a number of disadvantages: it requires the maximum priority of the threads using the mutex to be known *a priori*; if the ceiling priority is too high it acts as a global lock disabling all scheduling and it is pessimistic, taking action to prevent the problem even when it does not arise.

A better solution is to use a priority inheritance protocol. Here, the priority of the thread that owns the mutex is boosted to equal that of the highest priority thread that is waiting for it. This technique does not require *a priori* knowledge of the priorities of the threads that are going to use the mutex, and the priority of the owning thread is only boosted when a higher priority thread is waiting. This reduces the effect on the scheduling of other threads, and is more optimistic than the priority ceiling protocol. A disadvantage of this mechanism is that the cost of each synchronisation call is increased since the inheritance protocol must be obeyed each time. Another approach to priority inversion is to recognise that relative thread priorities have been poorly chosen and thus the system in which it occurs is faulty. In this case the kernel needs the ability to detect when priority inversion has taken place, and to raise an exception when it occurs to aid debugging. Then this code is removed from the shipping version.

It is worth noting that all these approaches have overheads to a certain extent and programmers need to carefully identify a suitable approach for a high-performance reliable solution.

5.5.2 Interprocess Communication

This is one of the key issues related to data dependencies and network configuration of a parallel architecture. Consider communication of two concurrent processes, one writing data and one reading data. They must adopt some type of synchronisation so that the reader knows when the writer has completed and the writer knows that the reader is ready for more data. In practice, programming environments provide explicit communication mechanisms such as message passing. Threads share all global variables, which provide thread programmers plenty of opportunities for synchronisation. Multiple processes can use many inter-process communication mechanisms, for instance, sockets, shared memory, and messages.

5.5.3 Scheduling and Mapping

Scheduling and mapping provides a clear description of how the operations of each thread can be executed by the processors, the allocation of each variable of the program to memory, and the specification of a single flow of control, common to all processors. In practice, multiprogrammed operating systems handle dynamic scheduling for the CPU. By switching the CPU among processes, the operating system can make the computer more productive. The operating system uses its scheduler to select from the pool of tasks that are ready to run. In a sense, the scheduler synchronises access of a task to a shared resource, for instance, CPUs of the system. However, neither the multithreaded version of a program nor the multiprocess version imposes any specific scheduling requirements on its tasks (Silverschatz and Galvin, 1998).

5.5.4 Granularity for Parallel Thread Execution

The granularity of an algorithm and in turn the program to execute a problem, is one of the important criteria for determining an appropriate parallel architecture. In general, the granularity for parallel program execution can be classified as:

- Statement level parallelism: offers an opportunity for fine-grained execution.
- Procedure level parallelism: offers an opportunity for medium-grained execution.
- Program level parallelism: offers an opportunity for coarse-grained execution.

In practice, algorithm granularity, number of PEs, size of the physical memory, and type of interconnection network are different from one implementation to another. Several factors may affect the granularity of a solution for a particular problem. For instance, if a solution requires communication with a central process for each small piece of computation, then the solution is fine grained. However, if the computation requires communication only at the end of each step, depending on how data are assigned to each process of a program, this approach can range from fine to coarse grained. In some computational problems, individual statements can each become a separate process. Another possibility is for procedures to be assigned to processes or for processes to represent whole programs.

5.5.5 Sharing Process Resources

Sharing process resources among the threads is another key issue for concurrent real-time multi-threading. All threads in multithreaded concurrency models share the resources of the process in which they exist. Independent processes and, in turn, processes which do not have any data and control dependencies share nothing. Threads share such process resources as global variables and file descriptors. If one thread changes the value of any such resource, the change will be evident to any other thread in the process. The sharing of process resources among threads is one of the major performance advantages, as well as one of the most difficult programming aspects of the multithreaded programming model. Having this notion available to all threads in the same memory facilitates communication between threads. However, at the same time, results from unexpected and unwanted variable changes are more likely to occur.

5.6 Case Study

In this section, a flexible beam simulation algorithm is used to study and explore real-time concurrent multithread implementation issues in a high-performance multiprocessor domain. A dual processor based computing domain comprising of two Intel processors (1 GHz) is considered to demonstrate the critical issues for concurrent multithreading in real-time computing. This investigation also addresses the issues of synchronisation, task granularity, scheduling and mapping, data and control dependencies and interprocess communication. Finally, a comparison of the results of the implementations is made on the basis of real-time performance to detail the advantges of system design incorporating fast processing multithreading techniques in a multiprocessor domain.

To investigate the parallel multithread implementation of the flexible beam simulation algorithm, a cantilever beam of length $L = 0.635\,\mathrm{m}$, mass $m = 0.0375\,\mathrm{kg}$ and $\lambda = 0.3629$ was considered. The beam simulation algorithm is developed using finite difference techniques, by discretising the corresponding governing dynamic equation of the beam in time and along the beam length. In this manner the beam is divided into a number of equal length sections and a difference equation describing the deflection of the beam is developed for each section. Further details and derivation of the algorithm are given in Chapter 6. For an adequate accuracy to be achieved in representing the first few (dominant) modes of vibration of the beam, the beam was divided into at least 20 segments.

A series of experiments was conducted to explore concurrent multithreading in a multiprocessor computing domain. In the first set of experiments (Section 5.6.1) Algorithm 1 (as discussed in Chapter 6) was used with simple generic mutual exclusion based synchronisation to explore the real-time performance. In the second set of experiments (Section 5.6.2), a number of synchronisation mechanisms with I/O options (for visualisation and file storing) were designed and tested for a number of algorithms to explore the real-time computing performance of multithreading in a multiprocessor domain.

5.6.1 Concurrent Multithreading for Flexible Beam Simulation

As mentioned above, this section presents a set of experiments to explore the real-time computing requirement for a flexible beam simulation Algorithm 1, as described in Chapter 6. The basic mutual exclusion principle was used for thread synchronisation to explore the advantages of real-time computation within the multi-processor computing domain.

Figure 5.1 shows the execution time for a single thread and two threads with synchronisation. It is noticed that due to synchronisation, interprocess communication overheads and other operating system overheads, a significant level of performance degradation was observed for two threads compared to a single thread. In an ideal situation, it would be expected that the execution time to implement the algorithm with two threads within a dual-processor domain would be half that of the single thread. However, the actual performance achieved with two threads is much worse than with a single thread and this trend continues with increase in the number of iterations.

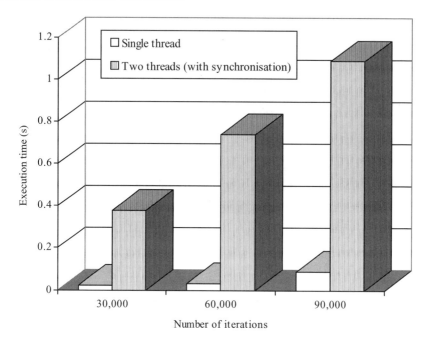

Figure 5.1. Execution time for a single thread and two threads with synchronisation (20 beam segments)

Figure 5.2 shows the execution time for a single thread and for two threads without synchronisation (considering that there are no data dependencies between the two threads) for a similar computation as discussed above. The basic objective to compute without synchronisation is to address the impact of the synchronisation overhead.   It is evident that the speedup for two threads in a dual-processor environment is double that of a single thread, and this remains almost consistent with increase in the number of iterations.

Figure 5.3 shows the execution time for two threads with and without synchronisation. It is further observed that the impact of synchronisation for inter-process resource sharing due to data dependencies is extremely high. This, in turn, causes a significant level of performance degradation with two-threads execution in a dual-processor domain when implementing the simulation algorithm.

To explore the impact of synchronisation and communication overhead further, two threads with 20, 40, 60 and 80 segments were also investigated. Figure 5.4 shows the execution time for 20, 40, 60 and 80 segments when implementing the algorithm with a single thread and two threads. It is noted that the overhead for 20, 40, 60 and 80 segments are consistent and of a similar level, although the computing load varies for the different numbers of segments.

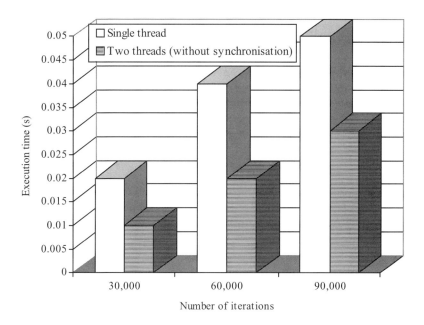

Figure 5.2. Execution time for a single thread and two threads without synchronisation (20 beam segments)

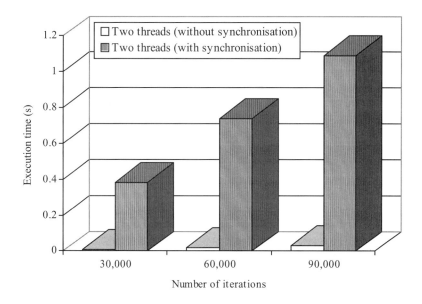

Figure 5.3. Execution time when implementing two threads with and without synchronisation (20 beam segments)

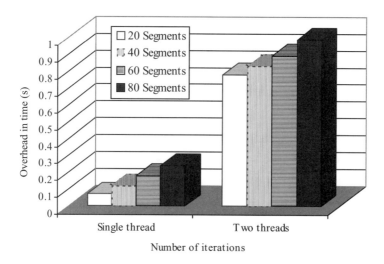

Figure 5.4. Execution time when implementing single and two threads with 20, 40, 60 and 80 beam segments

Figure 5.5 shows the execution time when implementing the simulation algorithm for different numbers of threads and iterations. It is noted that for a specific number of iterations, the execution time increases with increase in the number of threads. This increment is almost linear, which demonstrates that owing to synchronisation overheads, the execution time increases with increase in the number of threads. Although it is expected that the execution time would reduce with an increase in the number of threads, in practice, this is not so. This is due to the impact of the overheads of synchronisation, interprocess communication, operating system scheduling with an increasing number of threads.

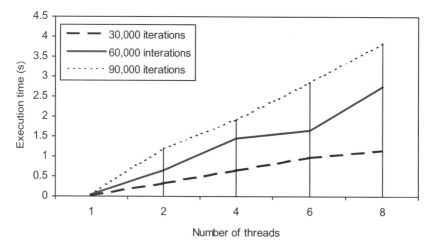

Figure 5.5. Execution time when implementing the algorithm with various numbers of threads

Table 5.1 shows the multithread execution time with and without synchronisation. It is noticed that a significant level of performance enhancement is achieved for two threads compared to a single thread without synchronisation. However, this improvement is not consistent or evident with further increase in the number of threads. This could be due to limitation of the number of processors. This demonstrates that for a computing domain comprising two processors, two threads might give the best mapping rather than further levels of granularity. Moreover, operating system overhead, scheduling and mapping could cause further levels of performance degradation with increase in the number of threads.

Table 5.1. Multithread execution time (s) when implementing the beam simulation algorithm with and without synchronisation (Sync.)

Number of threads	30,000 Iterations		60,000 Iterations		90,000 Iterations	
	With Sync.	Without Sync.	With Sync.	Without Sync.	With Sync.	Without Sync.
1	0.02	0.02	0.04	0.04	0.05	0.05
2	0.33	0.01	0.66	0.02	1.20	0.02
4	0.66	0.02	1.44	0.05	1.92	0.07
6	0.98	0.03	1.64	0.06	2.86	0.07
8	1.16	0.03	2.75	0.06	3.86	0.09

Table 5.2 shows the communication overhead and task granularity for 90,000 iterations. It is noticed that the level of communication overhead is significantly higher than the actual computation time, which in turn, reduces the task granularity ratio. This further demonstrates the impact of overhead due to a significant level of thread dependencies, and in turn, the inherent data dependencies of the beam simulation algorithm.

Table 5.2. Communication overhead and task granularity (for 90,000 iterations)

Number of threads	Execution time with synchronisation	Execution time without synchronisation (R)	Communication overhead (C)	Task granularity (R/C)
1	0.05	0.05	-----	---
2	1.20	0.02	1.18	0.017
4	1.92	0.07	1.85	0.038
6	2.86	0.07	2.79	0.025
8	3.86	0.09	3.77	0.024

5.6.2 Concurrent Multithreading with Visual and File I/O

This section introduces multithreading with visual and file I/O having different types of synchronisation mechanism, to evaluate comparative performances of the different algorithms discussed in Chapter 6. The first set of experiments with multithreaded programs consider the flexible beam simulation Algorithm 5 with 20 segments. The calculation of beam positions and their visual representation were performed on two distinct threads. Four different approaches employing Algorithm 5 in a multiprocessor domain were explored. The investigation also includes the suitability and reliability of synchronisation mechanisms in the algorithms.

Figure 5.6 shows the single-thread implementation model, considered as a reference, with which the performance of multithreaded programs can be compared.

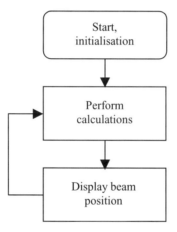

Figure 5.6. Single-thread implementation model

The first experiment incorporated a multithreaded implementation of the algorithm and utilised the pthread_join method to synchronise the two existing threads. The thread that performed the visualisation element was responsible for creation of the calculation thread during each iteration. This thread was suspended until the calculation thread terminated. The resulting program was considered as multithreaded Algorithm 1 (MTalg1), as shown in Figure 5.7.

The next multithreaded program (MTalg2) as shown in Figure 5.8 was developed with the aim of improving performance. The overhead of creating and terminating threads at each iteration was removed. This method used a much more efficient synchronisation mechanism between threads, which existed throughout the entire simulation. A mutex variable was introduced, which was locked by the calculation thread and unlocked by the visualisation thread.

Finally, experiments on a multithreaded implementation without any form of thread-level synchronisation were conducted. Assuming that the calculation thread performed significantly faster than the visualisation thread, the stopping criterion decision was made on the basis of process termination. In one mechanism this is after completion of the requested number of iterations (MTalg3b) as shown in Figure 5.9 and in the other mechanism it is after completion of the visualisation thread for the given number of iterations (MTalg3a) as shown in Figure 5.10. Algorithm MTalg3b had a very short runtime compared to all other implementations. The MTalg3a algorithm imposed a much higher load on the CPU, because a significantly large number of (not displayed) calculations were performed. Neither of these algorithms were synchronised to provide an opportunity for smooth visualisation.

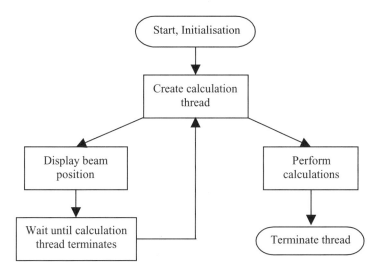

Figure 5.7. Algorithm MTalg1 with synchronisation

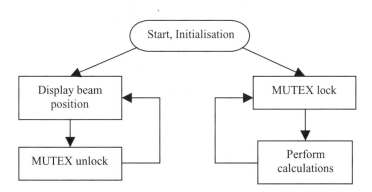

Figure 5.8. Algorithm MTalg2 with synchronisation

In order to interpret the results, it is important to recognise the existence of a strong I/O bottleneck. As a result of this bottleneck the total program duration was almost unaffected by the choice of thread implementation, as shown in Figure 5.11.

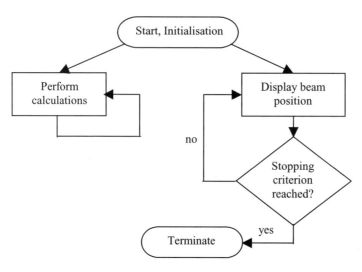

Figure 5.9. Algorithm MTalg3a with synchronisation

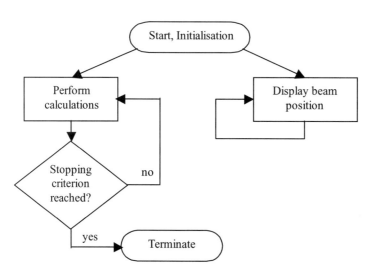

Figure 5.10. Algorithm MTalg3b with synchronisation

It is noted from the first experiment that the poor performance of the visualisation component might prevent valuable data from being extracted from further experiments. No genuine difference in total program execution time among the selected implementations can be observed in Figure 5.11. However, the execution time of the processes carried out by the CPU was recorded independently, and the performance diversity of the implementations became apparent. It is noted that, of the overall system resources, typically only a small percentage (around 15%) was used by the process itself, but a much greater proportion was used by the computing domain, to which the application sent the graphical data in order to display it. The obvious disparity between the results in Figure 5.11 and Figure 5.12 was due to the severe bottleneck at the visual I/O in the computing domain.

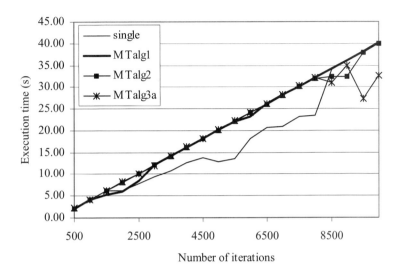

Figure 5.11. Execution time to implement the algorithms with visual I/O

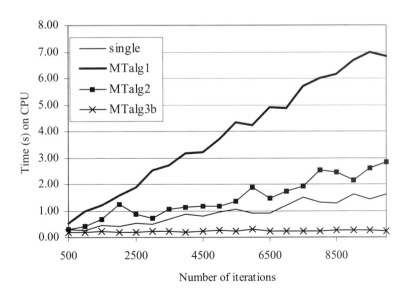

Figure 5.12. Execution time to implement the algorithms with visual I/O done by the CPU

It is well understood that the I/O routine will always perform slower than the calculation thread. The mutex variable was employed to ensure that the calculation thread would not proceed if the I/O bound thread did not complete its task. No precautions for the considerably less likely event of the I/O thread finishing ahead of the calculation thread was taken.

The algorithm MTalg2 was redesigned for a greater level of reliability by introducing a second mutex variable, as shown in Figure 5.13. The rest of the experiments in this section were conducted based on this modified Algorithm MTalg2.

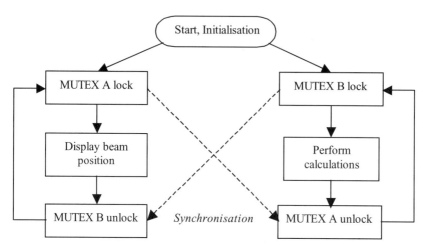

Figure 5.13. Modified algorithm MTalg2 with synchronisation

5.6.3 Performance Comparison with Concurrent Programs

In response to the bottleneck at the visual I/O encountered previously, three experiments were conducted and the results are presented here. The programs in each of these experiments were set to perform different amounts of visual I/O. Moreover, the results of single-threaded programs were plotted next to the program completion time of multithreaded programs to assess the benefits of concurrently executing different tasks of the algorithm.

In the first experiment a program was developed to display the beam on the screen after the end of each calculation thread. In the second experiment the beam position was calculated and displayed only each second. For the sake of brevity, only the programs for display at each iteration, every 4th iteration and every 16th calculation step are illustrated.

In these experiments 10,000 iterations were considered with Algorithm 5 and Algorithm 7, due to their relatively good performance and their flexibility regarding changes of parameters. The performance of these algorithms with a visualisation component and optionally with an additional file I/O component is graphically presented in subsequent figures. In each case, single- and multithreaded implementations of the algorithms are shown in comparison with each other. It is noted that the calculation thread of Algorithm 5 was in theory 33% more CPU intensive at each step. The main comparisons and subsequent analyses had to be made on different implementations of each algorithm individually.

Figures 5.14 and 5.15 show the program execution time of the algorithms considered with heavy visual I/O. A new beam is visualised on the screen at every iteration of the calculation thread. This situation was encountered in previous experiments. All algorithms and implementations thereof yield approximately the same performance behaviour, because of the severe bottleneck in the overall system.

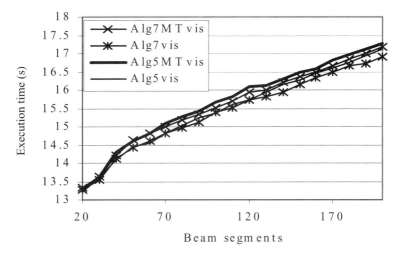

Figure 5.14. Execution times when implementing the algorithms with heavy visual I/O

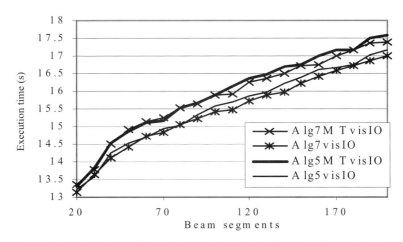

Figure 5.15. Execution times when implementing the algorithms with heavy visual and file I/O

It is however possible to identify that multithreaded implementations of the algorithms perform slightly worse than the simpler single-threaded versions. Figure 5.16 displays CPU utilisation required to perform the tasks. This utilisation is the sum of user-level CPU utilisation and system-level utilisation, which is performed by the operating system on behalf of the program. Compared to Figure 5.14, a significant disparity between the time the program needs to complete and the time it actually spends on the CPUs was noted. The CPU utilisation of these programs with heavy visual I/O in combination with file I/O produces a very similar picture, revealing no further useful information.

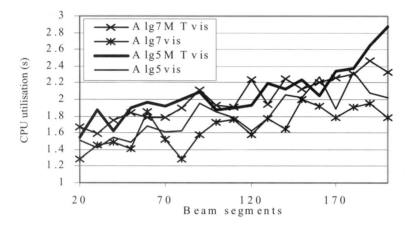

Figure 5.16. Time spent in CPU for heavy visual I/O

The next set of experiments considered programs with less heavy demand required by the visualisation component. The programs in this investigation were required to display the beam on the screen at every 4th calculation only.

Figure 5.17 demonstrates that even with a more moderate amount of visual I/O, the programs still perform very similarly. The bottleneck in the computing domain is still severe. As shown in Figure 5.18, adding a third thread for file I/O did not significantly change the results. It is however possible to make out the typical S-curve transitions in performance on segment sizes between 20 and 50. These transitions were characteristic of the selected algorithms on the target hardware used.

The last in this series of experiments was performed with an even smaller amount of visualisation I/O demanded by the programs. The data from every 16th calculation cycle is displayed as a beam on the screen.

As noted in Figure 5.19 the same performance behaviour identified in the previous Figures 5.14 and 5.17 was shown. However, the program execution times decreased towards the levels utilised by the CPU, largely eliminating the effects of the visualisation I/O bottleneck. The multithreaded versions of the programs still performed worse than the programs with a single thread, probably due to the overhead of the dual-mutex synchronisation method.

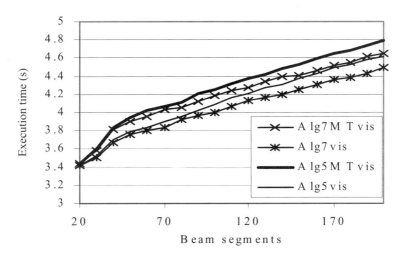

Figure 5.17. Program completion time with moderate visual I/O

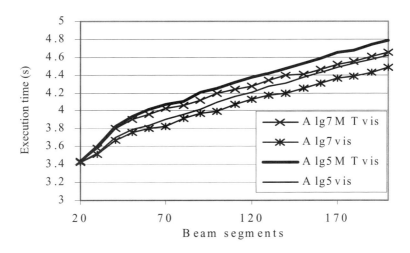

Figure 5.18. Execution time when implementing algorithms with moderate visual and file I/O

The CPU utilisation (user and system) by programs with only a light visual I/O bottleneck is fairly constant as the number of segments is increased. The increase in time required for the calculation thread is not clearly reflected by the data shown in Figure 5.20. This is rather surprising, especially as the early performance S-curve transitions could not be identified in Figure 5.20.

A description of the CPU utilisation in percentage terms is given in Figures 5.21 and 5.22. Experimental results for programs with a quarter or a sixteenth of the visual I/O in the original version are presented as vis 1/4 and vis 1/16, respectively.

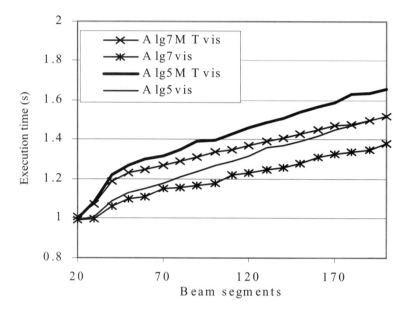

Figure 5.19. Execution time when implementing algorithms with light visual I/O

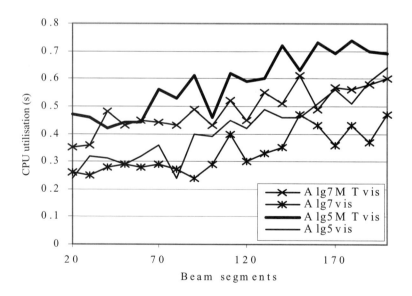

Figure 5.20. CPU time for light visual I/O

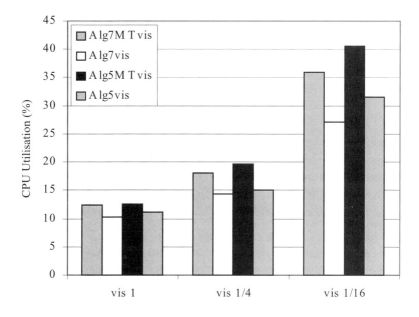

Figure 5.21. CPU utilisation time for visual I/O

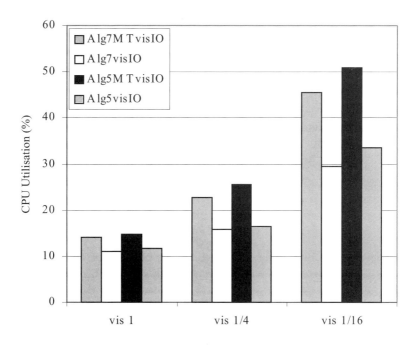

Figure 5.22. CPU utilisation for combined visual and file I/O

The multithreaded programs undoubtedly produced higher CPU utilisation at runtime. However, considering that two or more threads are available, occupying the dual-processor system resources with a theoretical maximum CPU utilisation of 200%, the resulting performances were not very impressive.

5.7 Summary

This chapter has described parallel programming and multithreading for real-time computing. A brief outline of parallel programming languages and program models has been given to develop a basic understanding of the topic. With an introduction to multithreading a generic form of parallel processing using multithreading in a multiprocessor computing domain has also been described. Moreover, real-time multithreading issues for multiprocessor computing domains have been discussed and demonstrated through a case study.

In the case study, an investigation into concurrent multithreading in multiprocessor domains for real-time high-performance computing, emphasising thread synchronisation and interprocess communication, has been presented, and demonstrated with the real-time implementation of a flexible beam simulation algorithm. A number of mechanisms for thread synchronisation for the flexible beam simulation algorithm have been designed and verified. A further investigation into the task granularity due to interprocess communication overhead and data dependency of the algorithm has been carried out. It has been demonstrated that data dependency is one of the critical issues for multithreading in real-time high-performance computing. It has also been shown that for enhanced performance, the number of threads is limited by the number of available processors in the computing domain for the application. It has been found that fine-grain threading with high levels of data dependency could cause significant performance degradation. In this particular application, increasing the number of threads increases the synchronisation overhead, which, in turn, relates with interprocess dependencies, operating system overhead for runtime scheduling and memory management.

5.8 Exercises

1. Describe the differences between multiprocessing and parallel processing.

2. Indicate the most important characteristics that a parallel programming language should possess. Describe procedural and non-procedural programming languages for parallel processing.

3. What is the main weakness of a sequential programming model for parallel processing? Describe the disparity between a synchronous and asynchronous parallel programming model.

4. Discuss concurrent multithreading for multiprocessor computing. What are the key issues of real-time concurrent multithreading for multiprocessor domains?

5. What are the most common methods for synchronisation? Describe the differences between semaphores and mutual exclusion for thread synchronisation.

6. What are the different levels of granularity for parallel thread execution? Give an example for each level of parallelism.

7. It is noted in the case study that concurrent multithreading is not suitable for real-time signal processing and control. Indicate the major issues behind this?

8. Design the following algorithm for concurrent multithreading using the synchronisation mechanism shown Figure 5.13. Implement the algorithm on a multiprocessor-based computing domain and evaluate the performances of multithread compared to single thread.

$$P(x+3,t) = P(x+2,t) + b^2 P(x+1,t) - c^2 P(x,t)$$

where, P is an array, initial value of the array can be considered as zero, b and c are small constants, t is the number of iterations and the value of x is 80.

6. Algorithm Analysis and Design

6.1 Objectives

- To introduce the algorithms used throughout this book.
- To explore data dependencies and control structures of the algorithms.
- To demonstrate algorithmic design issues.
- To explore the inherent parallelism in the algorithms.
- To demonstrate the real-time implementation issues of the algorithms.

6.2   Introduction

This chapter is concerned with the study of algorithms and analysis of their inherent characteristics. Analysis and design of algorithms emerged as a new scientific subject during the 1960s and has been quickly established as one of the most active fields of study, and an important part of computer and systems science. The reason for this sudden interest in the study of algorithms is not difficult to trace, as fast and successful development of digital computers and their use in different areas of human activity have led to the construction of a variety of computer algorithms.

In many cases, analysis of algorithms leads to the revelation of completely new algorithms that are even faster than algorithms previously known. The main goal in algorithmic analysis is to gain sufficient understanding of the relative merits of complicated algorithms that useful information is provided to someone undertaking an actual computation.

In practice, more than one algorithm exists for solving a specific problem. Depending on its formulation, each can be evaluated numerically in different ways. As computer arithmetic is of finite accuracy different results can evolve, depending on the algorithm used and the way it is evaluated. The choice of the best algorithm for a given problem and for a specific computer is a difficult task and depends on many factors, including data and control dependencies, granularity and regularity of the algorithm (Thoeni, 1994).

Data dependency is a key issue in algorithm analysis for real-time high-performance computing. This includes both sequential and parallel real-time processing. During the implementation of an algorithm, data dependency between two blocks or two statements requires memory access time. In practice, further dependencies imply further access time, or more interprocessor communication in the case of parallel processing. This, accordingly, degrades the real-time performance. Thus, it is essential to study and analyse data dependencies in an algorithm for real-time implementation. Such a study should address essential questions as: how to reduce block or statement dependencies? how to reduce memory access time? and, what is the impact of data dependencies on real-time interprocessor communication?

Complex and computationally demanding algorithms are often employed to satisfy the performance requirements of modern signal processing and control systems. This requires proper partitioning and mapping of the algorithm to be implemented. A considerable amount of research work has been devoted to the exploration of the computing processes, partitioning and mapping of algorithms for implementation in different computing domains (Agrawal *et al.*, 1986; Akl, 1989; Baxter *et al.*, 1994; Crummey *et al.*, 1994; Cvetanovic, 1987; Hwang, 1993; Khokhar *et al.*, 1993; Maguire, 1991; Megson, 1992; Nation *et al.*, 1993; Nocetti and Fleming, 1991; Sing *et al.*, 1993). A number of strategies have emerged through these investigations for task partitioning and task to processor mapping in an attempt to achieve better performance. Important issues that have to be considered in parallel implementation of an algorithm include:

- The amount of parallelism inherent in the algorithm.
- The methodology of decomposing a problem into subproblems.
- The method applied to allocate these subproblems to processors.
- The grain size of the subproblem for execution on each processor.
- The possibility of overlapping processing with communication.
- Runtime memory management requirements for the algorithm code developed.

In this chapter an investigation into the characteristics and the design methodology of an algorithm within the framework of uniprocessor and multiprocessor computing domains for real-time implementation is presented. The algorithms considered include an adaptive active vibration control (AVC) algorithm, the LMS and the recursive least squares (RLS) adaptive filtering algorithms, the correlation algorithm, the FFT algorithm and the simulation algorithm for a flexible manipulator system. The algorithms are described with systematic computing methods and code development for implementation on uniprocessor and multiprocessor computing systems. The methodology also explores the partitioning and task allocation process for multiprocessor implementations. Both homogeneous and heterogeneous mapping structures and computing demands are considered for better performance with matching and mismatching of algorithms and architectures.

6.3 Data and Control Dependencies of Algorithms

Dependencies determine how one instruction depends on another. Dependencies are a property of computer programs, which in some cases limit the performance of the program in terms of efficiency and execution time. The main classes of dependencies between two blocks or two statements or two variables are

- Data dependence.
- Control dependence.

The key importance of data dependencies are the following:

i.   dependence indicates the possibility of a hazard;
ii. it determines the order in which results must be calculated; and
iii. dependence sets an upper bound on how much parallelism can possibly be exploited.

A dependence can be overcome in two different ways:

1. maintaining the dependence but by appropriate source code scheduling;
2. eliminating a dependence by transforming the code.

Detection of parallelism in an application involves finding sets of computations that can be performed simultaneously. The approach to parallelism is based on the study of data dependencies. The presence of dependence between two computations implies that they cannot be performed in parallel. In general, the fewer the dependencies, the greater the parallelism.

Many algorithms have regular data dependencies, that is dependencies that repeat throughout the set of computations in the algorithm. For such algorithms dependencies can be concisely described mathematically and can be manipulated easily. However, there are algorithms for which dependencies vary from one computation to another and these algorithms are more difficult to analyse. When two or more algorithms have similar dependencies, it means that they exhibit similar parallel properties.

A control dependency, on the other hand, determines the ordering of an instruction '*i*' with respect to a branch instruction so that the instruction '*i*' is executed in correct program order and only when it should be. Note that control dependency is preserved by two properties in a simple sequential computing:

a. instructions executed in program order, ensuring that an instruction that occurs before a branch is executed before the branch;
b. the detection of control ensures that an instruction that is control dependent on a branch is not executed until the branch direction is known.

To identify algorithmic parallelism, it is essential to explore the basic structural features of the algorithm as dictated by its data and control dependencies. The presence of dependencies indicates complexity in an algorithm and, in turn, communication overhead in a parallel processing context.

6.4 Granularity and Regularity

Granularity and regularity are two important issues of algorithm analysis and design for high-performance sequential and parallel computing. In particular, the study of parallelism includes interprocessor communication, issues of granularity of the algorithm and of the hardware and regularity of the algorithm. Hardware granularity, as described in Chapter 4, is defined as the ratio of computational performance over the communication performance of each processor within the architecture:

$$\text{Hardware granularity} = \frac{\text{Runtime length of a task}}{\text{Communication overhead}} = \frac{R}{C}$$

where R is the actual computation time of a task and C is the amount of time due to communication overhead during execution of the corresponding task. When R/C is very small, it is unprofitable to use parallelism. When R/C is very large, parallelism is potentially profitable. A characteristic of fine-grain processors is that they have fast interprocessor communication, and can therefore tolerate small task sizes and still maintain a satisfactorily high R/C ratio. However, medium-grain or course-grain processors with slower interprocessor communication will produce correspondingly smaller R/C ratios if their task sizes are also small (Hossain, 1995).

Task granularity can be defined as the ratio of computational demand (or time required to execute the task without any communication overhead) of a task over the communication demand (actual communication time) during execution of the corresponding task. Typically a high compute/communication ratio is desirable. The concept of task granularity can also be viewed in terms of compute time per task. When this is large, it is a coarse-grain task implementation. When it is small, it is a fine-grain task implementation. Although coarse granularity may ignore potential parallelism, partitioning a problem into the finest possible granularity does not necessarily lead to the fastest solution, as maximum parallelism also has maximum overhead, particularly due to increased data dependencies and, in turn, communication requirements. Therefore, when partitioning the algorithm into subtasks and distributing these across the PEs, it is essential to choose an algorithm granularity that balances useful parallel computation against communication and other overheads (Nocetti and Fleming, 1991).

Generally, algorithms can be classified on the basis of their characteristics as regular, irregular and mixed (combined regular and irregular). Regularity is a term used to describe the degree of uniformity in the execution thread of the computation. Many algorithms can be expressed by matrix computations. This leads to the so-called regular iterative type of algorithms due to their very regular structure. In addition to their own characteristics, the regularity or the irregularity of algorithms also depends on how the code is developed for implementation. On some occasions, a regular algorithm could result from specific coding of a task of an irregular nature. Computing hardware, on the other hand, also possesses its own speciality leading to performance variation of the architecture with algorithms. For example, an architecture achieving higher performance for a regular algorithm may achieve lower performance for an irregular algorithm. A vector processor, for instance, in principle, may achieve better performance for a regular algorithm as compared to an irregular algorithm. Thus, it is essential to explore the characteristics of the algorithm, hardware and coding style for computation. This will lead to an exploration of the matching and mismatching of the algorithm requirements and the hardware resources.

6.5 Analysis of Algorithms with Computational Aspects

6.5.1 Adaptive Active Vibration Control

Three algorithms, namely, simulation, control and identification are involved in the adaptive AVC algorithm. The adaptive AVC algorithm is considered within a ＊ single-input single-output (SISO) structure. The real-time implementation of the algorithm is investigated within the beam simulation algorithm characterising a cantilever beam in transverse motion. The process of implementation of the algorithms is described and discussed in this Section with due consideration of the issues of computational granularity, partitioning and mapping.

Beam Simulation Algorithm
Consider a cantilever beam system with a force $U(x,t)$ applied at a distance x from its fixed (clamped) end at time t. This results in a deflection $y(x,t)$ of the beam from its stationary position at the point where the force has been applied. In this manner, the governing dynamic equation of the beam is given by:

$$\mu^2 \frac{\partial^4 y(x,t)}{\partial x^4} + \frac{\partial^2 y(x,t)}{\partial t^2} = \frac{1}{m} U(x,t) \tag{6.1}$$

where μ is a beam constant and m is the mass of the beam. By discretising the beam in time and length using central finite difference (FD) methods a discrete approximation to Equation 6.1 can be obtained as (Virk and Kourmoulis, 1988):

$$Y_{k+1} = -Y_{k-1} - \lambda^2 SY_k + \frac{(\Delta t)^2}{m} U(x,t) \tag{6.2}$$

where, $\lambda^2 = \left[(\Delta t)^2 / (\Delta x)^4\right] \mu^2$ with Δt and Δx representing the step sizes in time and along the beam respectively,

$$
Y_{j+1} = \begin{bmatrix} y_{1,j+1} \\ y_{2,j+1} \\ \vdots \\ y_{n,j+1} \end{bmatrix}, \quad
Y_j = \begin{bmatrix} y_{1,j} \\ y_{2,j} \\ \vdots \\ y_{n,j} \end{bmatrix}, \quad
Y_{j-1} = \begin{bmatrix} y_{1,j-1} \\ y_{2,j-1} \\ \vdots \\ y_{n,j-1} \end{bmatrix}
$$

and S is a penta-diagonal matrix, given (for $n = 20$, say) as

$$
S = \begin{bmatrix}
a & -4 & 1 & 0 & 0 & 0 & \cdots & \cdots & 0 \\
-4 & b & -4 & 1 & 0 & 0 & \cdots & \cdots & 0 \\
1 & -4 & b & -4 & 1 & 0 & \cdots & \cdots & 0 \\
0 & 1 & -4 & b & -4 & 1 & \cdots & \cdots & 0 \\
\cdots & \cdots & \cdots & \cdots & \cdots & \cdots & \cdots & \cdots & \cdots \\
\cdots & \cdots & \cdots & \cdots & \cdots & \cdots & \cdots & \cdots & \cdots \\
\cdots & \cdots & \cdots & \cdots & 1 & -4 & b & -4 & 1 \\
\cdots & \cdots & \cdots & \cdots & 0 & 1 & -4 & c & -2 \\
\cdots & \cdots & \cdots & \cdots & 0 & 0 & 2 & -4 & d
\end{bmatrix}
$$

where, $a = 7 - \dfrac{7}{\lambda^2}$, $b = 6 - \dfrac{2}{\lambda^2}$, $c = 5 - \dfrac{2}{\lambda^2}$ and $d = 2 - \dfrac{2}{\lambda^2}$. Equation 6.2 is the required relation for the simulation algorithm, characterising the behaviour of the cantilever beam system, which can be implemented easily on a digital computer. For the algorithm to be stable it is required that the iterative scheme described in Equation 6.2, for each grid point, converges to a solution. It has been shown that a necessary and sufficient condition for stability satisfying this convergence requirement is given by $0 < \lambda^2 \le 0.25$ (Kourmoulis, 1990).

This is a regular type matrix based vector algorithm, involving mainly the computation of the deflection of 20 segments (or above) of the beam (fixed-free). If this is a fine-grain algorithm, it can be split into 20 different equal components. However, note that to comput the deflection of each segment information on the deflection of the previous and the next two segments is required. This implies that, if the algorithm is implemented as fine-grain or even in coarse-grain form, it will demand a heavy communication overhead among the individual components. The nature of the algorithm also implies that a vector processor could be suitable for its implementation.

In the case of uniprocessor based architectures, the algorithm can be implemented as a sequential process. In the case of multiprocessor based architectures, on the other hand, the algorithm can be partitioned for equal load distribution among the processors. With heterogeneous architectures the algorithm can be partitioned according to the capabilities of the PEs, by allocating a heavier load to the high-performance processor rather than the low-performance processor. The allocation could be made after obtaining the performance of the PEs of the corresponding parallel architecture. Further details of scheduling and mapping issues of the algorithm were provided in Chapter 3.

Control Algorithm

A schematic diagram of an AVC structure is shown in Figure 6.1. A detection sensor detects the unwanted (primary) disturbance. This is processed by a controller to generate a cancelling (secondary, control) signal to achieve cancellation at the observation point. The objective in Figure 6.1 is to achieve total (optimum) vibration suppression at the observation point. Synthesising the controller on the basis of this objective yields (Tokhi and Hossain, 1994)

$$C = \left[1 - \frac{Q_1}{Q_0} \right]^{-1} \tag{6.3}$$

where, Q_0 and Q_1 represent the equivalent transfer functions of the system (with input at the detector and output at the observer) when the secondary source is *off* and *on* respectively. Equation 6.3 is the required controller design rule, which can easily be implemented on-line on a digital processor. This leads to a self-tuning AVC algorithm comprising the processes of identification and control. The process of identification involves obtaining Q_0 and Q_1 using a suitable system identification algorithm. An RLS parameter estimation algorithm, described later is used here to estimate Q_0 and Q_1 in the discrete-time domain in parametric form. The process of control, on the other hand, involves designing the controller according to Equation 6.3 and implementing this in real-time.

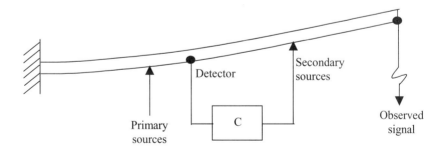

Figure 6.1. Active vibration control structure

The process of calculating the controller parameters uses a set of design rules based on Equation 6.3. Let the system models Q_0 and Q_1 be described by

$$Q_0 = \frac{b_{00} + b_{01}z^{-1} + b_{02}z^{-2}}{1 + a_{01}z^{-1} + a_{02}z^{-2}} \quad , \quad Q_1 = \frac{b_{10} + b_{11}z^{-1} + b_{12}z^{-2}}{1 + a_{11}z^{-1} + a_{12}z^{-2}} \tag{6.4}$$

Substituting for Q_0 and Q_1 from Equation 6.4 into Equation 6.3 and simplifying yields the required SISO controller transfer function as

$$C = \frac{b_{C0} + b_{C1}z^{-1} + b_{C2}z^{-2} + b_{C3}z^{-3} + b_{C4}z^{-4}}{1 + a_{C1}z^{-1} + a_{C2}z^{-2} + a_{C3}z^{-3} + a_{C4}z^{-4}} \tag{6.5}$$

where

$$
\begin{aligned}
b_{C0}(b_{00} - b_{10}) &= b_{00} \\
b_{C1}(b_{00} - b_{10}) &= b_{01} + b_{00}a_{11} \\
b_{C2}(b_{00} - b_{10}) &= b_{02} + b_{01}a_{11} + b_{00}a_{12} \\
b_{C3}(b_{00} - b_{10}) &= b_{02}a_{11} + b_{01}a_{12} \\
b_{C4}(b_{00} - b_{10}) &= b_{02}a_{12} \\
a_{C1}(b_{00} - b_{10}) &= b_{01} + b_{00}a_{11} - b_{10}a_{01} - b_{11} \\
a_{C2}(b_{00} - b_{10}) &= b_{02} + b_{01}a_{11} + b_{00}a_{12} - b_{10}a_{02} - b_{11}a_{01} - b_{12} \\
a_{C3}(b_{00} - b_{10}) &= b_{02}a_{11} + b_{01}a_{12} - b_{11}a_{02} - b_{12}a_{01} \\
a_{C4}(b_{00} - b_{10}) &= b_{02}a_{12} - b_{12}a_{02}
\end{aligned}
\tag{6.6}
$$

This gives the set of design rules for calculation of the required controller parameters.

Once the controller parameters are estimated, the required controller can be implemented by inserting the values into the controller transfer function, Equation 6.5. The controller response can then be obtained either using frequency-domain methods, by replacing z^{-1} (inverse-shift operator) by $\exp(-j\omega T)$ or in the time domain using the corresponding difference equation

$$y(n) = \sum_{i=0}^{4} b_{Ci}u(n-i) - \sum_{j=1}^{4} a_{Cj}y(n-j) \tag{6.7}$$

where $u(n)$ and $y(n)$ correspond to the discrete input and output signals of the controller.

Note that in implementing Equation 6.7 within the beam simulation environment, the simulation algorithm becomes an integral part of the process. Thus, the control algorithm consists of the combined implementation of Equation 6.7 and the beam simulation algorithm.

The control algorithm, as outlined, is essentially composed of the beam simulation algorithm and the difference Equation 6.7. Thus, in uniprocessor based architectures the algorithm can be implemented as a sequential process. In two-processor based architectures, on the other hand, the algorithm can be partitioned for the beam simulation part in a similar manner as in the case of the simulation algorithm with one of the PEs additionally carrying out calculation of the control signal according to Equation 6.7. From a practical point of view, Equation 6.7 has similar computing load as one beam segment. Therefore, the task distribution of this algorithm within homogeneous and heterogeneous architectures should be similar to that of the simulation algorithm.

Identification Algorithm
The identification algorithm is described here as the process of estimating parameters of the required SISO controller characteristics. It consists of the processes of estimating the system models Q_0 and Q_1 and the controller design calculation. The RLS algorithm is used here for estimation of parameters of the system models Q_0 and Q_1. This is based on the well known least squares method.

Let an unknown plant with input $u(n)$ and output $y(n)$ be described by a discrete linear model of order m as

$$y(n) = b_0 u(n) + b_1 u(n) + ... + b_m u(n-m) - a_1 y(n-1) - ... - a_m y(n-m)$$

or

$$y(n) = \Psi(n)\Theta(n) \tag{6.8}$$

where Θ is the model parameter vector and Ψ, known as the observation matrix, is a row vector of the measured input/output signals. In this manner, the RLS estimation process at time step k is described by (Tokhi and Leitch, 1992)

$$\varepsilon(k) = \Psi(k)\Theta(k-1) - y(k)$$

$$\Theta(k) = \Theta(k-1) - P(k-1)\Psi^T(k)\left[1 + \Psi(k)P(k-1)\Psi^T(k)\right]^{-1}\varepsilon(k) \tag{6.9}$$

$$P(k) = P(k-1) - P(k-1)\Psi^T(k)\left[1 + \Psi(k)P(k-1)\Psi^T(k)\right]^{-1}\Psi(k)P(k-1)$$

where $P(k)$ is the covariance matrix. Thus, the RLS estimation process implements and executes the relations in Equation 6.9 in the order given. The performance of the estimator can be monitored by observing the parameter set at each iteration. Once convergence has been achieved the routine can be stopped. Convergence is determined by the magnitude of the modelling error $\varepsilon(k)$ or when the estimated set of parameters reaches a steady level.

The RLS algorithm used for estimation of parameters of the system models Q_0 and Q_1 can be outlined as:

i.   compute the error $\varepsilon(k)$ using

$$\varepsilon(k) = \Psi(k)\Theta(k) - y(k)$$

ii. update the parameter vector using

$$\Theta(k) = \Theta(k-1) - \mathbf{P}(k-1)\Psi^T(k)\left[1 + \Psi(k)\mathbf{P}(k-1)\Psi^T(k)\right]^{-1}\varepsilon(k)$$

iii. update the covariance matrix using

$$\mathbf{P}(k) = \mathbf{P}(k-1) - \mathbf{P}(k-1)\Psi^T(k)\left[1 + \Psi(k)\mathbf{P}(k-1)\Psi^T(k)\right]^{-1}\Psi(k)\mathbf{P}(k-1)$$

Once a steady set of parameters for both models has been obtained, it can be used to obtain the parameters of the controller. Thus, in the case of uniprocessor based architectures, the algorithm was implemented as a sequential process. In the case of two-processor based architectures, on the other hand, the algorithm can be partitioned so that the load incurred estimating parameters of Q_0 and Q_1 is equally distributed among the two PEs, with one of the PEs further carrying out calculation of the controller parameters. Moreover, in this process, limited communication (due to parameters of Q_0 and Q_1) for calculation of controller parameters is required between the two PEs implementing the algorithm. Load distribution of this algorithm on more than two-processor-based homogeneous or heterogeneous architecture would be complex due to the impact of huge data and control dependencies. Therefore, unexpected performance degradation would be observed. However, the data and control dependencies of the algorithm have to be looked at closely to schedule the algorithm within a homogeneous architecture with more than two processors or in a heterogeneous architecture.

The identification algorithm described above is a mixed type. In fact, the irregular part of the algorithm is more influential than the regular matrix part. Thus, this algorithm could be suitable for a processor with features of a more irregular computing nature.

The processes of estimating parameters of the models Q_0 and Q_1 are similar. This implies that the same program coding can be used to estimate the parameters of Q_0 and Q_1. Following a process of initialisation, which involves setting the parameter vector $\Theta(0)$ to 0, say, and the covariance matrix $\mathbf{P}(0) = \rho I$, with ρ as a constant of the order of 500 to 1000, the RLS algorithm is executed as described above. The performance of the estimator can be monitored by observing the parameter set at each iteration. Once a steady set of parameter values have been obtained the routine can be stopped. To evaluate the performance of the RLS parameter estimation algorithm, the flexible beam simulation environment was utilised. The beam was excited by a step disturbance and the response monitored at two grid-points. The deflection at the detection point of the beam (nearest point of the disturbance) was considered as the plant input. On the other hand, the deflection at the observation point (target vibration control point) was considered as the plant output. The characteristics of the 'plant' thus defined were estimated using a second-order model.

6.5.2 Adaptive Filtering

The contamination of a signal of interest by unwanted, often larger amplitude, signal or noise is a problem often encountered in many applications. When the signal and noise occupy fixed and separate frequency bands, conventional linear filters with fixed coefficients are normally used to extract the signal. However, there are many instances where there is a spectral overlap between the signal and the noise or the band occupied by the noise is unknown and varies with time. In such cases it is necessary for the filter characteristics to vary, or be adapted with changing signal statistics.

The schematic diagram of an adaptive filter is shown in Figure 6.2. The adaptive filter operates by estimating the statistics of the incoming signal and adjusting its own characteristics so as to minimise some cost function. This cost function may be derived in a number of ways depending on the intended application. The cost function is normally derived through the use of a secondary signal or conditioning input as shown in Figure 6.2. This secondary signal input $y(n)$ may be defined as the desired output of the filter, in which case the task of the adaptive algorithm is to adjust the weights in the programmable filter device to minimise the difference or error $e(n)$ between the filter output $\hat{y}(n)$ and the desired signal $y(n)$. These types of adaptive filters are frequently used to recover signals from channels with time-varying characteristics (Crown and Grant, 1985). Two widely used adaptive filtering algorithms, namely the LMS and RLS, are considered here.

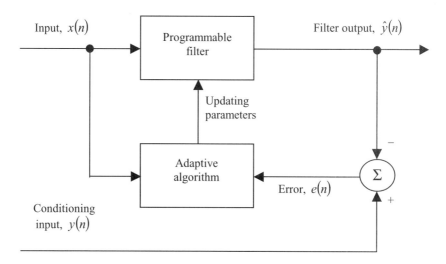

Figure 6.2. Schematic diagram of an adaptive filter

LMS Filter

The LMS algorithm is one of the most successful adaptive algorithms developed by Widrow and co-workers (Widrow *et al.*, 1975). It is based on the steepest descent method where the weight vector is updated according to

$$\mathbf{W}_{k+1} = \mathbf{W}_k - 2e_k \mu \mathbf{X}_k \qquad (6.10)$$

where $\mathbf{W}_k$ and $\mathbf{X}_k$ are the weight and the input signal vectors at time step k respectively, μ is a constant controlling the stability and rate of convergence and e_k is the error given by

$$e_k = y_k - \mathbf{W}_k^T \mathbf{X}_k \qquad (6.11)$$

where y_k is the current contaminated signal sample. It is clear from the above that the LMS algorithm does not require prior knowledge of the signal statistics. The weights obtained by the LMS algorithm are not only estimates, but these are adjusted so that the filter learns the characteristics of the signals leading to a convergence of the weights. The condition for convergence is given by

$$0 < \mu > 1/\lambda_{max} \qquad (6.12)$$

where λ_{max} is the maximum eigenvalue of the input data covariance matrix.

Following an initialisation process which involves setting the weights $w(i)$, $i = 0,1,...,N-1$, to arbitrary fixed values, zero say, and the parameter μ to a suitable value, the LMS algorithm consists of implementing the steps below at each sample time k :

i.   measure the next input sample and compute the filter output using

$$\overset{\Lambda}{y}_k = \sum_{i=0}^{N-1} w_k(i) x_{k-i}$$

ii.   compute the error e_k using

$$e_k = y_k - \hat{y}_k$$

iii.   update the filter weights using

$$w_{k+1}(i) = w_k(i) + 2\mu e_k x_{k-i}$$

It is easily understood from the above that the amount of computation involved in step i is similar to that in step iii whereas step ii involves only a minor subtraction operation. Thus, in the case of a homogeneous architecture incorporating two processors, steps i and iii can each be allocated to individual processors with step ii given to either of the two processors. In the case of a heterogeneous architecture of two processors, however, step ii can be allocated to the faster performing processor with steps i and iii each given to an individual processor. As observed, scheduling and mapping of the LMS algorithm in the case of more than three processors in either a homogeneous or heterogeneous architecture would be difficult because of its irregular nature of computation. Although, the three components of the algorithm can be distributed easily among processors within a three-processor architecture, the performance might be degraded due to mismatch of the load distribution in a homogeneous architecture. On the other hand, the algorithm can easily be distributed among processors within a heterogeneous architecture of three processors on the basis of performance of the processors and the computational load of different components of the algorithm. Therefore, possible performance enhancement could be expected in the case of a three-processor based heterogeneous architecture.

RLS Filter
The RLS algorithm is based on the well-known least squares method. An output signal $y(k)$ of the filter is measured at the discrete time k, in response to a set of input signals $x(k)$ (Tokhi and Leitch, 1992). The error variable is given by

$$\varepsilon(k) = \Psi(k)\Theta(k-1) - y(k) \tag{6.13}$$

where Θ and Ψ represent the parameter vector and the observation matrix of the filter respectively. These are given by

$$\Theta^{\mathrm{T}} = [\theta(1),\quad \theta(2),\quad ...,\quad \theta(m)]$$

$$\Psi = [\psi(1),\quad \psi(2),\quad ...,\quad \psi(m)]$$

where m represents the order and ψ the input sample of the filter. The new parameter vector is given by

$$\Theta(k) = \Theta(k-1) - \mathbf{P}(k-1)\Psi^{T}(k)[1 + \Psi(k)\mathbf{P}(k-1)\Psi^{T}(k)]^{-1}\varepsilon(k) \tag{6.14}$$

with $\mathbf{P}(k)$, representing the covariance matrix at time step k, given by

$$\mathbf{P}(k) = \mathbf{P}(k-1) - \mathbf{P}(k-1)\Psi^{T}(k)[1 + \Psi(k)\mathbf{P}(k-1)\Psi^{T}(k)]^{-1}\Psi(k)\mathbf{P}(k-1) \tag{6.15}$$

The performance of the filter can be monitored by observing the error variable $\varepsilon(k)$ at each iteration.

Following a process of initialisation which involves setting the parameter vector $\Theta(0)$ to 0, say, and the covariance matrix $\mathbf{P}(0) = \rho I$, with ρ as a constant of the order of 500 to 1000, the RLS algorithm consists of implementing the steps below at each time step k.

   i.   measure the next filter input sample and compute the filter output using

$$\hat{y}(k) = \Psi(k)\Theta(k-1)$$

   ii.   compute the error $\varepsilon(k)$ using

$$\varepsilon(k) = \Psi(k)\Theta(k-1) - y(k)$$

   iii.   update the parameter vector using

$$\Theta(k) = \Theta(k-1) - \mathbf{P}(k-1)\Psi^T(k)\left[1 + \Psi(k)\mathbf{P}(k-1)\Psi^T(k)\right]^{-1}\varepsilon(k)$$

   iv.   update the covariance matrix using

$$\mathbf{P}(k) = \mathbf{P}(k-1) - \mathbf{P}(k-1)\Psi^T(k)\left[1 + \Psi(k)\mathbf{P}(k-1)\Psi^T(k)\right]^{-1}\Psi(k)\mathbf{P}(k-1)$$

It is clear from the above that the computational load involved in steps *iii* and *iv* is more than that in steps *i* and *ii*. Thus, to minimise the communication overhead, in the case of a homogeneous architecture of two processors, steps *i* and *ii* can be allocated to one processor, and steps *iii* and *iv* to the other processor. In the case of a heterogeneous architecture, however, steps *i* and *ii* can be allocated to the slower performing processor and steps *iii* and *iv* to the faster performing processor. Due to significant data and control dependencies of the algorithm, it would be difficult to achieve better performance using more than two processors with either homogeneous or heterogeneous architectures. Moreover, variation of computational load in different sections of the algorithm is noticeably high and therefore, it would be difficult to achieve proper scheduling and mapping for better performance.

Both the LMS and RLS algorithms as described above are irregular, although LMS is more irregular than RLS. Thus, for both algorithms, a processor with an irregular computing characteristic will be more suitable for real-time implementation.

6.5.3 Spectral Analysis

For spectral analysis, correlation and FFT algorithms are considered in this investigation. Correlation is a measure of similarity between two waveforms. It is a method of time-domain analysis that is particularly useful for detecting periodic signals buried in noise, for establishing coherence between random signals and a source. Applications are found in many engineering fields such as radar, radio astronomy, medical, nuclear and acoustical research (Proakis and Manolakis, 1988). The FFT, on the other hand, plays an important role in many DSP applications including linear filtering and spectral analysis. A major reason for its importance is the existence of efficient algorithms (Ludeman, 1986). A divide-and-conquer approach is used in this investigation to implement the FFT algorithm.

Correlation Algorithm
Cross-correlation is a measure of the similarity between two waveforms. Consider two signal sequences $x(n)$ and $y(n)$ each of finite energy. The cross-correlation between $x(n)$ and $y(n)$ is a sequence $r_{xy}(l)$ defined as

$$r_{xy}(l) = \sum_{n=-\alpha}^{\alpha} x(n)y(n-l); \; l = 0,\pm1,\pm2,... \tag{6.16}$$

or, equivalently, as

$$r_{xy}(l) = \sum_{n=-\alpha}^{\alpha} x(n+l)y(n); \; l = 0,\pm1,\pm2,... \tag{6.17}$$

The index l is the (time) shift (or lag) parameter and the subscripts xy on the cross-correlation sequence $r_{xy}(l)$ indicate the sequences being correlated. The order of the subscripts, with x preceding y, indicates the direction in which one sequence is shifted, relative to the other. To elaborate, in Equation 6.16, the sequence $x(n)$ remains unshifted and $y(n)$ is shifted by l units of time, to the right for l positive and to the left for l negative. Equivalently, in Equation 6.17, the sequence $y(n)$ is left unshifted and $x(n)$ is shifted by l units in time, to the left for l positive and to the right for l negative. But shifting $x(n)$ to the left by l units relative to $y(n)$ is equivalent to shifting $y(n)$ to the right by l units relative to $x(n)$. Hence the computations in Equations 6.16 and 6.17 yield identical cross-correlation sequences (Proakis and Manolakis, 1988).

The correlation algorithm yielding the cross-correlation sequence between two sequences $x(n)$ and $y(n)$ consists of the steps below.

    i.   collect the two sequences $x(n)$ and $y(n)$;

ii. calculate the intermediate value for positive correlation for s samples, using

$$r_{xy}(l) = \sum_{n=0}^{s-l} x(n)y(n+l); \ l = 0,1,2,...$$

iii. calculate the intermediate value for negative correlation, using

$$r_{xy}(-l) = \sum_{n=0}^{s-l} x(n+l)y(n); \ l = 0,1,2,...$$

To compute the correlation between the two signals, steps *ii* and *iii* require the same amount of calculation with no communication between them. Thus, in the case of a homogeneous architecture of two PEs steps *ii* and *iii* each can be allocated to a PE. In the case of a heterogeneous architecture of two PEs, on the other hand, the full load of step *ii* and part of step *iii* can be allocated to the higher performing processor and the rest of step *iii* to the lower performing processor. It is clear from the above that, irrespective of the structure of task allocation, no communication will be needed between the two PEs. It is noted that due to data and control dependencies, decomposition of the load for allocation to more than two processors in an architecture will result in performance degradation.

FFT Algorithm
A discrete-time signal $x(n)$ of duration N can be expressed as a weighted sum of complex exponential sequences. Since sinusoidal sequences are unique only at discrete frequencies from 0 to 2π, the expansion contains only a finite number of complex exponentials as

$$x(n) = \frac{1}{N} \sum_{k=0}^{N-1} X(k)e^{jk\omega_0 n} \quad \text{for } 0 \le n \le N-1 \tag{6.18}$$

where, the coefficients of expansion $X(k)$ and the fundamental discrete frequency ω_0 are given by

$$X(k) = \sum_{n=0}^{N-1} x(n)e^{-jk\omega_0 n} \ ; \omega_0 = 2\pi/N \quad \text{for } 0 \le k \le N-1 \tag{6.19}$$

Equations 6.18 and 6.19 constitute the discrete Fourier transform (DFT) relations. Equation 6.19 is sometimes written in the following equivalent form:

$$X(k) = \sum_{n=0}^{N-1} x(n)W_N^{kn} \tag{6.20}$$

where, W_N is defined as

$$W_N = e^{-j2\pi/N} \tag{6.21}$$

Since $x(n)$ may be complex $X(k)$ can be written as

$$
\begin{aligned}
X(k) = \sum_{n=0}^{N-1} &\{(\mathrm{Re}[x(n)]\,\mathrm{Re}[W_N^{kn}] - \mathrm{Im}[x(n)]\,\mathrm{Im}[W_N^{kn}]) \\
&+ j(\mathrm{Re}[x(n)]\,\mathrm{Re}[W_N^{kn}] + \mathrm{Im}[x(n)]\,\mathrm{Im}[W_N^{kn}])\}
\end{aligned}
\tag{6.22}
$$

From Equation 6.22 it is clear that for each value of k, the direct computation of $X(k)$ requires $4N$ real multiplications and $4N-2$ real additions. Since $X(k)$ must be computed for N different values of k, the direct computation of the DFT of a sequence $x(n)$ requires $4N^2$ real multiplications and $N(4N-1)$ real additions or, alternatively, N^2 complex multiplications and $N(N-1)$ complex additions. In addition to the multiplications and additions called for by Equation 6.22, the implementation requires provision for storing and accessing the input sequence values $x(n)$, values of the coefficients W_N^{kn} and intermediate results (Proakis and Manolakis, 1988; Ludeman, 1986).

Computationally efficient DFT algorithms are known collectively as FFT algorithms. The divide-and-conquer approach is one of the FFT algorithms. It is based on decomposition of an N-point DFT into successively smaller DFTs. In this case, for the computation of an N-point DFT, N can be considered as a product of two integers, that is, $N = LM$. The assumption that N is not prime is not restrictive since any sequence can be padded with zeros to ensure a factorisation of this form. The sequence $x(n)$, $0 \leq n \leq N-1$, can, thus, be stored as a two-dimensional array of $l \times m$, where $0 \leq l \leq L-1$ and $0 \leq m \leq M-1$. Note that l is the row index and m is the column index. Thus, the sequence $x(n)$ can be stored in a rectangular array in a variety of ways each of which depends on the mapping of index n to the indices (l, m). Suppose the selected mapping is, $n = lM + m$; this leads to an arrangement in which the first row consists of the first M elements of $x(n)$, the second row consists of the next M elements $x(n)$, and so on. On the other hand, the mapping $n = mL + l$ stores the first L elements of $x(n)$ in the first column, the next L elements in the second column, and so on (Proakis and Manolakis, 1988).

A similar arrangement can be used to store the computed DFT values. In particular, the mapping is from the index k to a pair of indices (p, q), where $0 \leq p \leq L-1$ and $0 \leq q \leq M-1$. If the selected mapping is, $k = pM + q$, the DFT is stored on a row-wise basis, where the first row contains the first M elements of $X(k)$, the second row contains the next set of M elements and so on. On the other hand, the mapping, $k = qL + p$ results in a column, the second set of L elements are stored in the second column, and so on.

Suppose that $x(n)$ is mapped onto the rectangular array $x(l, m)$ and $X(k)$ is mapped onto a corresponding rectangular array $X(p, q)$, then the DFT can be expressed as a double sum over the elements of the rectangular array multiplied by the corresponding phase factors. A column-wise mapping for $x(n)$ and a row-wise mapping for the DFT can be adopted (Ifeachor and Jervis, 1993). Thus,

$$X(p,q) = \sum_{m=0}^{M-1}\sum_{l=0}^{L-1} x(l,m)W_N^{(Mp+q)(mL+l)} \tag{6.23}$$

But,

$$W_N^{(Mp+q)(mL+l)} = W_N^{MLmp}\,W_N^{mLq}\,W_N^{Mpl}\,W_N^{lq}$$

However, $W_N^{Nmp} = 1, W_N^{mqL} = W_{N/L}^{mq} = W_M^{mq}$, and $W_N^{Mpl} = W_{N/M}^{pl} = W_L^{pl}$. With these simplifications, Equation 6.23 can be expressed as

$$X(p,q) = \sum_{l=0}^{L-1}\left\{W_N^{lq}\left[\sum_{m=0}^{M-1} x(l,m)W_M^{mq}\right]\right\}W_L^{lp} \tag{6.24}$$

Equation 6.24 involves the computation of DFTs of lengths M and L. Thus, the FFT algorithm can be described as:

i. compute the M point DFTs using

$$F(l,q) = \sum_{m=0}^{M-1} x(l,m)W_M^{mq}, \ 0 \le q \le M-1 \ \text{ for each row } l = 0,1,...,L-1$$

ii. compute a new rectangular array $G(l, q)$ as

$$G(l,q) = W_N^{lq}F(l,q), \quad 0 \le l \le L-1, 0 \le q \le M-1$$

iii. compute the L-point DFTs using

$$X(p,q) = \sum_{l=0}^{L-1} G(l,q)W_L^{lp}\ , \ \text{ for each column } q = 0,1,...,M-1 \ \text{ of array}$$
$$G(l,q)$$

On the surface it may appear that the computational procedure outlined above is more complex than the direct computation of DFT. However, as noted, step i

involves the computation of L DFTs, each of M points. Hence this step requires LM^2 complex multiplications and $LM(M-1)$ complex additions. Step ii requires LM complex multiplications. Finally, step iii in the computation requires LM^2 complex multiplications and $LM(L-1)$ complex additions. Therefore, the computational complexity is

- complex multiplications: $N(M+L+1)$;
- complex additions: $N(M+L-2)$;

where $N = LM$. Thus, the number of multiplications has been reduced from N^2 to $N(M+L+1)$ and the number of additions has been reduced from $N(N-1)$ to $N(M+L-2)$. For example, consider $N = 1000$, and select L and M as 2 and 500, respectively. Then, by direct computation the number of multiplications will be 1,000,000, but by the divide-and-conquer method it is 503000. This represents a reduction by approximately a factor of 2. The number of additions is also reduced by a factor of about 2.

The FFT is a regular matrix-based algorithm. Thus, a vector processor (for instance i860) can perform better than other processor types in implementing this algorithm. To compute the FFT of a time series on a multiprocessor architecture, a communication overhead is incurred. The amount of computation required in steps i and iii are similar. In contrast, step ii requires a very small amount of computation compared to steps i and iii. Thus, in the case of a homogeneous architecture of two processors, steps i and iii can each be allocated to a PE, with step ii to any one of the PEs. In contrast, with a heterogeneous architecture of two PEs, steps $i+ii$ or $ii+iii$ can be allocated to the high-performance processor and the remaining i or iii to the low-performance processor. With the above method, scaling and scheduling of the FFT algorithm within more than two processors is relatively easy compared to many other signal processing algorithms, however, it would be more suitable for massively parallel systolic array type architectures.

Both correlation and FFT algorithms are regular types. In fact, the FFT algorithm structure is more regular matrix based than the correlation algorithm. Thus, a vector processor may be more suitable for implementing these algorithms in real time.

6.5.4 Flexible Manipulator System

Real-time simulation of a complex flexible manipulator algorithm is very demanding. Although the structure of the algorithm is similar to that of the flexible beam system, in terms of computing and communication demand it is different. Thus, this algorithm is considered to investigate such computing and communication demands. A schematic representation of a single-link flexible manipulator is shown in Figure 6.3. A control torque τ is applied at the pinned end (hub) of the arm by an actuator motor. θ represents the hub angle, POQ is the original co-ordinate system (stationary coordinate) while $P'OQ'$ is the coordinate system after an angular rotation θ (moving coordinate). I_h is the inertia at the hub, I_p is the inertia associated with payload M_p at the end-point and u is the flexible displacement (deflection) of a point at a distance x from the hub. The dynamic equation of the flexible manipulator, considered as an Euler–Bernoulli beam equation, can be expressed as (Azad, 1994):

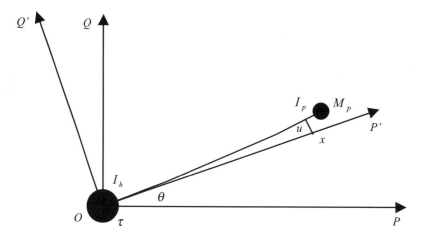

Figure 6.3. Schematic representation of the flexible manipulator system

$$\rho \frac{\partial^2 y(x,t)}{\partial t^2} + EI \frac{\partial^4 y(x,t)}{\partial x^4} = \tau(x,t) \qquad (6.25)$$

where $y(x,t)$ is the displacement (deflection) of the manipulator at a distance x from the hub at time t, ρ is the density per unit length of the manipulator material, E is Young's modulus, I is the second moment of inertia, and $\tau(x,t)$ is the applied torque. The product EI represents the flexural rigidity of the manipulator. The corresponding boundary conditions at the hub end are given as:

$$y(0,t) = 0$$
$$I_h \frac{\partial^3 y(0,t)}{\partial t^2 \partial x} - EI \frac{\partial^2 y(0,t)}{\partial x^2} = \tau(t) \qquad (6.26)$$

Similarly, the boundary conditions at the end-point of the manipulator are given by

$$M_p \frac{\partial^2 y(L,t)}{\partial t^2} - EI \frac{\partial^3 y(L,t)}{\partial x^3} = 0$$

$$I_p \frac{\partial^3 y(L,t)}{\partial t^2 \partial x} + EI \frac{\partial^2 y(L,t)}{\partial x^2} = 0$$

(6.27)

where, L is the length of the manipulator. The initial conditions along the t coordinate are described as:

$$y(0,t) = 0 \quad \text{and} \quad \frac{\partial y(x,0)}{\partial x} = 0$$

(6.28)

The above describes the state of behaviour of the flexible manipulator system, which allows construction of a simulation platform for testing and evaluation of various controller designs.

To solve the partial differential equation (PDE) in Equation 6.25 the FD method (Lapidus, 1982) is utilised. The manipulator length and movement time are each divided into a suitable number of sections of equal length represented by Δx ($x = i\Delta x$) and Δt ($t = j\Delta t$) respectively. A difference equation corresponding to each point of the grid is developed. The displacement $y_{i,j+1}$ at each time instant can then be written as (Azad, 1994):

$$y_{i,j+1} = -c\left[y_{i-2,j} + y_{i+2,j}\right] + b\left[y_{i-1,j} + y_{i+1,j}\right] + ay_{i,j} - y_{i,j-1} + \frac{\Delta t^2}{\rho}\tau(i,j)$$

(6.29)

where, $a = 2 - \dfrac{6\Delta t^2 EI}{\rho \Delta x^4}$, $b = \dfrac{4\Delta t^2 EI}{\rho \Delta x^4}$ and $c = \dfrac{\Delta t^2 EI}{\rho \Delta x^4}$. Equation 6.29 gives the displacement of section i of the manipulator at time step $j+1$. Using matrix notation, Equation 6.30 can be written as:

$$y_{i,j+1} = Ay_{i,j} - y_{i,j-1} + BF$$

(6.30)

where

$$y_{i,j+1} = \begin{bmatrix} y_{1,j+1} \\ y_{2,j+1} \\ \vdots \\ y_{n,j+1} \end{bmatrix}, \quad y_{i,j} = \begin{bmatrix} y_{1,j} \\ y_{2,j} \\ \vdots \\ y_{n,j} \end{bmatrix}, \quad y_{i,j-1} = \begin{bmatrix} y_{1,j-1} \\ y_{2,j-1} \\ \vdots \\ y_{n,j-1} \end{bmatrix}$$

$$A = \begin{bmatrix} m_1 & m_2 & m_3 & 0 & 0 & \cdots & 0 & 0 & 0 & 0 & 0 \\ b & a & -b & -c & 0 & \cdots & 0 & 0 & 0 & 0 & 0 \\ -c & b & a & b & -c & \cdots & 0 & 0 & 0 & 0 & 0 \\ \ddots & \ddots & \ddots & \ddots & \ddots & \ddots & \ddots & \ddots & \ddots & \ddots & \ddots \\ 0 & 0 & 0 & 0 & 0 & \cdots & -c & b & a & b & -c \\ 0 & 0 & 0 & 0 & 0 & \cdots & 0 & m_{11} & m_{12} & m_{13} & m_{14} \\ 0 & 0 & 0 & 0 & 0 & \cdots & 0 & m_{21} & m_{22} & m_{23} & m_{24} \end{bmatrix}$$

$$F = \begin{bmatrix} \tau(i,j) \\ 0 \\ \vdots \\ 0 \end{bmatrix}, \quad B = \frac{\Delta t^2}{\rho}$$

Equation 6.30 is the general solution of the PDE giving the displacement of section i of the manipulator at time step $j+1$.

It follows from Equation 6.29 that, to obtain the displacements $y_{1,j+1}$, $y_{n-1,j+1}$ and $y_{n,j+1}$ the displacements of the fictitious points $y_{-1,j}$, $y_{n+1,j}$ and $y_{n+2,j}$ are required. Estimation of these displacements is based on the boundary and initial conditions related to the dynamic equation of the flexible manipulator, which in turn determine the values of m_1 to m_3 and m_{11} to m_{24} in matrix A of Equation 6.30. In this manner, any change in a boundary or an initial condition will affect the elemental values of matrix A. Moreover, during the calculation process, it is important to provide some means of checking the boundary conditions at each time step, so that the matrix A is updated, upon detecting any change. In some cases all calculations for the fictitious points are included within the calculation loop during each iteration. This increases the computational burden on the processor. To overcome this problem, two assumptions are made:

a. once the system is in operation there will be no change in the conditions at the hub of the manipulator;
b. during an operation, a change in matrix A will occur only when there is a change in the payload. Moreover, this payload change will affect only eight elements of the matrix. Assuming the hub inertia fixed then unnecessary repetitive calculations of the displacement for a certain movement at each iteration can be avoided by the payload checking facility.

Algorithm stability can be examined by ensuring that the iterative scheme described in Equation 6.29 converges to a solution. The necessary and sufficient condition for stability satisfying this convergence requirement is given by (Azad, 1994).

$$0 \leq c \leq \frac{1}{4}$$

Thus, for the algorithm to be stable the parameter c is required to satisfy the above equation.

The simulation algorithm for the flexible manipulator system, as described above, is essentially given by Equation 6.30. This is very similar to the beam simulation algorithm, i.e., a regular matrix based vector algorithm, involving mainly the computation of displacement of n segments of the manipulator arm. Thus, a similar strategy can be adopted in implementing the algorithm on uni-processor architectures and in partitioning and mapping the algorithm for parallel implementation.

6.6 Case Study

In this section a set of exercises demonstrating the concepts discussed earlier in this chapter are presented. The beam simulation algorithm is considered as a sample example.

6.6.1 Algorithm Design

The beam simulation algorithm is of regular iterative type. In implementing this algorithm on a sequential vector processor a performance better than with any other processor can be expected. The algorithm processes floating-point data, which is computed within a small iterative loop. Accordingly, the performance is further enhanced if the processor has internal data cache, instruction cache and/or a built-in maths co-processor.

The simulation algorithm in Equation 6.2 can be rewritten, to compute the deflection of segments 8 and 16, as in Figure 6.4, assuming no external force applied at these points.

```
y[8][8] ← −y[8][6] − lumsq*(y[6][7]− 4*y[7][7] +  b*y[8][7] −4*y[9][7]+y[10][7]);
y[16][16]← y[16][14]−lumsq*(y[14][15]−4*y[15][15]+b*y[16][15]−4*y[17][15]+y[18][15]);
```

Figure 6.4. Calculation of deflection of segments 8 and 6 (where, lumsq is lambda square)

It follows from the above that computation of deflection of a segment at time step t can be described as in Figure 6.5. It is also noted that computation of deflection of a particular segment is dependent on the deflection of six other segments. These heavy dependencies could be major causes of performance degradation in real-time sequential computing, due to memory access time. On the other hand, these dependencies might cause significant performance degradation in real-time parallel computing due to interprocessor communication overheads.

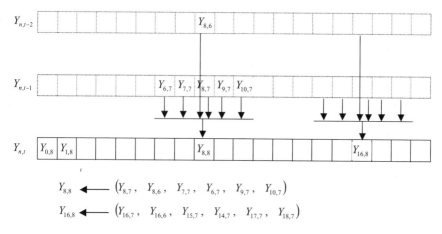

$$Y_{8,8} \longleftarrow \left(Y_{8,7}, \ Y_{8,6}, \ Y_{7,7}, \ Y_{6,7}, \ Y_{9,7}, \ Y_{10,7}\right)$$

$$Y_{16,8} \longleftarrow \left(Y_{16,7}, \ Y_{16,6}, \ Y_{15,7}, \ Y_{14,7}, \ Y_{17,7}, \ Y_{18,7}\right)$$

Figure 6.5. Data dependencies for computation of deflection of each segment

To explore this issue a number of design mechanisms for the beam simulation algorithm were developed to explore the real-time performance. Two of these algorithms are derived from previous work (Kabir *et al.*, 2000; Tokhi *et al.*, 1997b). These were subsequently investigated further. As a result, a total of seven designs of the simulation algorithm were developed and tested through a set of experiments. These designs are described below.

Algorithm 1: Shifting of Data Array
Algorithm 1 was adopted from a previously reported work (Tokhi *et al.*, 1997b). The FD algorithm for a flexible beam simulation was implemented through discrete approximation of the dynamic equation in time and along the beam length. The algorithm is listed in Figure 6.6. It is noted that complex matrix calculations are performed within an array of three elements, each representing information about the beam position at different instants of time. Subsequent to calculations, the memory pointer is shifted to the previous pointer in respect of time before the next iteration. This technique of shifting the pointer does not contribute to the calculation efforts and is thus a program overhead. Other algorithms were deployed to address this issue at further levels of investigation.

Note in Step 2 of Algorithm 1 in Figure 6.6 that at the end of every iteration, in shifting the deflection of the beam along the time scale, the value of the memory pointer y[1][1] is shifted to y[1][0] and the value of pointer y[1][2] is shifted to y[1][1]. This requires memory shifting twice, and thus four memory accesses. In the first memory access, the value is read from y[1][1]   and in the second memory access the value is shifted to y[1][0], in the third memory access, the value is read from y[1][2] and finally, in the fourth memory access the value is shifted to y[1][1]. Thus, if in this process the memory access time is T s, for a single iteration, with n segments of the beam, the total time for memory access will be $4nT$ s. Thus, for N iterations, the total memory access time will be $4nNT$ sec. As an example, if $T = 60 \, \text{ns}$, $n = 19$ and $N = 10,000$ then the total memory access time will be

$4 \times 19 \times 1000 \times 60 \times 10^{-9} = 0.0456 \, \text{s}$. This will be a comparably substantial percentage of the total execution time of the algorithm. Thus, it is necessary to eliminate this overhead due to extra memory access time. For this purpose, several design formulations of the algorithm are proposed to enhance the performance of the simulation algorithm.

```
Loop {
//Step 1
  y0[2]=-y0[0]-lamsq*(a*y0[1]-4*y1[1]+y2[1]);
  y1[2]=-y1[0]-lamsq*(-4*y0[1]+b*y1[1]-4*y2[1]+y3[1]);
  y2[2]=-y2[0]-lamsq*(y0[1]-4*y1[1]+b*y2[1]-4*y3[1]+y4[1]);
          :
  y17[2]=-y17[0]-lamsq*(y15[1]-4*y16[1]+b*y17[1]-4*y18[1]+y19[1]);
  y18[2]=-y18[0]-lamsq*(y16[1]-4*y17[1]+c*y18[1]-2*y19[1]);
  y19[2]=-y19[0]-lamsq*(2*y17[1]-4*y18[1]+d*y19[1]);
//Step 2
  // Shifting memory locations
  y0[0]=y0[1]; y0[1]=y0[2];
  y1[0]=y1[1]; y1[1]=y1[2];
          :
  y18[0]=y18[1]; y18[1]=y18[2];
  y19[0]=y19[1]; y19[1]=y19[2];
}
```

Fig. 6.6. Design of Algorithm 1

Algorithm 2: Array Rotation

Algorithm 2 incorporates design suggestions made by Kabir *et al.* (2000). A listing of Algorithm 2 is given in Figure 6.7. In this case, each loop calculates three sets of data. Instead of shifting the data of the memory pointer (that contains results) at the end of each loop, the most current data is directly recalculated and written into the memory pointer that contains the older set of data. Therefore, re-ordering of the array in Algorithm 1 is replaced by recalculation. The main objective of the design is to achieve better performance by reducing the dynamic memory allocation and, in turn, memory pointer shift operation. Thus, instead of using a single code block and data-shifting portion, as in Algorithm 1, to calculate the deflection, three code blocks, are used with the modified approach in Algorithm 2.

In Algorithm 2, the overhead of Algorithm 1 due to memory pointer shift operation is eliminated and every line of code is directed towards the simulation effort. A disadvantage of this approach is the increased number of instructions inside the main program loop and its susceptibility to coding errors. The latter issue can be successfully addressed by means of automatic source code generation.

```
Loop {
//Step 1
  y0[2]=-y0[0]-lamsq*(a*y0[1]-4*y1[1]+y2[1]);
  y1[2]=-y1[0]-lamsq*(-4*y0[1]+b*y1[1]-4*y2[1]+y3[1]);
  y2[2]=-y2[0]-lamsq*(y0[1]-4*y1[1]+b*y2[1]-4*y3[1]+y4[1]);
    :
  y17[2]=-y17[0]-lamsq*(y15[1]-4*y16[1]+b*y17[1]-4*y18[1]+y19[1]);
  y18[2]=-y18[0]-lamsq*(y16[1]-4*y17[1]+c*y18[1]-2*y19[1]);
  y19[2]=-y19[0]-lamsq*(2*y17[1]-4*y18[1]+d*y19[1]);
//Step 2
  y0[0]=-y0[1]-lamsq*(a*y0[2]-4*y1[2]+y2[2]);
  y1[0]=-y1[1]-lamsq*(-4*y0[2]+b*y1[2]-4*y2[2]+y3[2]);
  y2[0]=-y2[1]-lamsq*(y0[2]-4*y1[2]+b*y2[2]-4*y3[2]+y4[2]);
    :
  y17[0]=-y17[1]-lamsq*(y15[2]-4*y16[2]+b*y17[2]-4*y18[2]+y19[2]);
  y18[0]=-y18[1]-lamsq*(y16[2]-4*y17[2]+c*y18[2]-2*y19[2]);
  y19[0]=-y19[1]-lamsq*(2*y17[2]-4*y18[2]+d*y19[2]);
//Step 3
  y0[1]=-y0[2]-lamsq*(a*y0[0]-4*y1[0]+y2[0]);
  y1[1]=-y1[2]-lamsq*(-4*y0[0]+b*y1[0]-4*y2[0]+y3[0]);
  y2[1]=-y2[2]-lamsq*(y0[0]-4*y1[0]+b*y2[0]-4*y3[0]+y4[0]);
    :
  y17[1]=-y17[2]-lamsq*(y15[0]-4*y16[0]+b*y17[0]-4*y18[0]+y19[0]);
  y18[1]=-y18[2]-lamsq*(y16[0]-4*y17[0]+c*y18[0]-2*y19[0]);
  y19[1]=-y19[2]-lamsq*(2*y17[0]-4*y18[0]+d*y19[0]);
}
```

Figure 6.7. Design of Algorithm 2

Algorithm 3: Large Array and Less Frequent Shifting
In Algorithm 1 shifting of memory pointers was required within each iteration. Algorithm 3 was developed as an attempt to reduce the number of memory pointer shifting instructions and thereby to decrease program overhead. An array of 1000 elements was considered for each beam segment. This array size was chosen rather arbitrarily, but small enough to allow easy allocation of these monolithic memory blocks within typical hardware boundaries. Figure 6.8 shows how the array is utilised in Algorithm 3. Shifting occurs at the end of every thousandth iteration, rendering the overhead produced at this stage negligible. However, array positions are indirectly referenced through a variable, accessed at runtime which, in turn, leads to an overhead. Of far greater concern to program performance is the fact that large data structures need to be dealt with. Therefore, the internal data cache struggles to handle the large amount of data.

Algorithm 4: Nested Loops and Shifting
Algorithm 4 incorporates merely a minor modification of Algorithm 1, as shown in Figure 6.9. The aim in this algorithm is to contain the number of instructions inside the main loop, and thus, reduce the instruction size of the program. This was

accomplished by nesting secondary loops inside the main iterations. Complex substitutions need to be carried out to determine which matrix elements need to be referred to for performing the ongoing calculations. A moderate amount of overhead resulting from these necessary substitutions was anticipated. The benefits of this algorithm include quicker compilation, greater flexibility in respect of the number of segments (possibly changes at runtime) and a fixed number of program instructions in the main loop as segment sizes are increased. The likelihood of capacity misses in the instruction cache was significantly reduced.

```
Loop {
  for(j=0; j<1000; j++) {
    y0[j]=-y0[pj]-lamsq*(a*y0[ppj]-4*y1[ppj]+y2[ppj]);
    y1[j]=-y1[pj]-lamsq*(-4*y0[ppj]+b*y1[ppj]-4*y2[ppj]+y3[ppj]);
    y2[j]=-y2[pj]-lamsq*(y0[ppj]-4*y1[ppj]+b*y2[ppj]- 4*y3[ppj]+y4[ppj]);
              :
    y17[j]=-y17[pj]-lamsq*(y15[ppj]-4*y16[ppj]+b*y17[ppj]-4*y18[ppj]+y19[ppj]);
    y18[j]=-y18[pj]-lamsq*(y16[ppj]-4*y17[ppj]+c*y18[ppj]-2*y19[ppj]);
    y19[j]=-y19[pj]-lamsq*(2*y17[ppj]-4*y18[ppj]+d*y19[ppj]);
    pj++; ppj++;
  }
  // Shifting memory locations
  y0[0] = y0[998]; y0[1] = y0[999];
  y1[0] = y1[998]; y1[1] = y1[999];
       :
  y18[0] = y18[998]; y18[1] = y18[999];
  y19[0] = y19[998]; y19[1] = y19[999];
}
```

Figure 6.8. Design of Algorithm 3

```
Loop {
  y[0][2]=-y[0][0]-lamsq*(a*y[0][1]-4*y[1][1]+y[2][1]);
  y[1][2]=-y[1][0]-lamsq*(-4*y[0][1]+b*y[1][1]-4*y[2][1]+y[3][1]);
  for (i=2; i<18; i++){
    y[i][2]=-y[i][0]-lamsq*(y[i-2][1]-4*y[i-1][1]+b*y[i][1]-4*y[i+1][1]+y[i+2][1]);
  }
  y[18][2]=-y[18][0]-lamsq*(y[16][1]-4*y[17][1]+c*y[18][1]-2*y[19][1]);
  y[19][2]=-y[19][0]-lamsq*(2*y[17][1]-4*y[18][1]+d*y[19][1]);
  // Shifting memory locations
  for (i=0; i<20; i++) {
    y[i][0]=y[i][1]; y[i][1]=y[i][2];
  }
}
```

Figure 6.9. Design of Algorithm 4

Algorithm 5: Nested Loops and Array Rotation
Figure 6.10 shows a listing of Algorithm 5, in which the new methods of Algorithm 4 were applied with the concepts of Algorithm 2. Three distinct calculation runs are performed during each iteration, but instead of listing the instructions for each segment separately, nested loops are used to limit the amount of instructions (source code lines) in the main program loop. The benefits of employing this technique are identical with those listed in the description of Algorithm 4. However, it possesses the same disadvantage of overhead produced by the complex substitutions required.

```
Loop {
  // Step 1
  y[0][2]=-y[0][0]-lamsq*(a*y[0][1]-4*y[1][1]+y[2][1]);
  y[1][2]=-y[1][0]-lamsq*(-4*y[0][1]+b*y[1][1]-4*y[2][1]+y[3][1]);
  for (i=2; i<18; i++){
    y[i][2]=-y[i][0]-lamsq*(y[i-2][1]-4*y[i-1][1]+b*y[i][1]-4*y[i+1][1]+y[i+2][1]);
  }
  y[18][2]=-y[18][0]-lamsq*(y[16][1]-4*y[17][1]+c*y[18][1]-2*y[19][1]);
  y[19][2]=-y[19][0]-lamsq*(2*y[17][1]-4*y[18][1]+d*y[19][1]);
  // Step 2
  y[0][0]=-y[0][1]-lamsq*(a*y[0][2]-4*y[1][2]+y[2][2]);
  y[1][0]=-y[1][1]-lamsq*(-4*y[0][2]+b*y[1][2]-4*y[2][2]+y[3][2]);
  for (i=2; i<18; i++){
    y[i][0]=-y[i][1]-lamsq*(y[i-2][2]-4*y[i-1][2]+b*y[i][2]-4*y[i+1][2]+y[i+2][2]);
  }
  y[18][0]=-y[18][1]-lamsq*(y[16][2]-4*y[17][2]+c*y[18][2]-2*y[19][2]);
  y[19][0]=-y[19][1]-lamsq*(2*y[17][2]-4*y[18][2]+d*y[19][2]);
  // Step 3
  y[0][1]=-y[0][2]-lamsq*(a*y[0][0]-4*y[1][0]+y[2][0]);
  y[1][1]=-y[1][2]-lamsq*(-4*y[0][0]+b*y[1][0]-4*y[2][0]+y[3][0]);
  for (i=2; i<18; i++){
    y[i][1]=-y[i][2]-lamsq*(y[i-2][0]-4*y[i-1][0]+b*y[i][0]-4*y[i+1][0]+y[i+2][0]);
  }
  y[18][1]=-y[18][2]-lamsq*(y[16][0]-4*y[17][0]+c*y[18][0]-2*y[19][0]);
  y[19][1]=-y[19][2]-lamsq*(2*y[17][0]-4*y[18][0]+d*y[19][0]);
}
```

Figure 6.10. Design of Algorithm 5

Algorithm 6: Two-element Array Rotation
Algorithm 6 is shown in Figure 6.11. This makes use of the fact that access to the oldest time segment is only necessary during re-calculation of the same longitudinal beam segment. Hence, it can directly be overwritten with the new value as shown in Figure 6.12.

Figures 6.13 and 6.14 show simplified flow diagrams of Algorithm 2 and Algorithm 6, respectively. The conventional re-calculation algorithm in Figure 6.6 requires three memory segments in the time domain. In contrast, Algorithm 6 is optimised for the particular discrete mathematical approximation of the governing physical formula, exploiting the previously observed features.

It is noted that this particular algorithm is not suitable for applications for which the previous assumption doss not hold. This technique attributes to a major performance advantage over the conventional rotation method, in particular when the number of beam segments is increased.

```
Loop {
  // Step 1
  y0[0]=-y0[0]-lamsq*(a*y0[1]-4*y1[1]+y2[1]);
  y1[0]=-y1[0]-lamsq*(-4*y0[1]+b*y1[1]-4*y2[1]+y3[1]);
  y2[0]=-y2[0]-lamsq*(y0[1]-4*y1[1]+b*y2[1]-4*y3[1]+y4[1]);
         :
  y17[0]=-y17[0]-lamsq*(y15[1]-4*y16[1]+b*y17[1]-4*y18[1]+y19[1]);
  y18[0]=-y18[0]-lamsq*(y16[1]-4*y17[1]+c*y18[1]-2*y19[1]);
  y19[0]=-y19[0]-lamsq*(2*y17[1]-4*y18[1]+d*y19[1]);
  // Step 2
  y0[1]=-y0[1]-lamsq*(a*y0[0]-4*y1[0]+y2[0]);
  y1[1]=-y1[1]-lamsq*(-4*y0[0]+b*y1[0]-4*y2[0]+y3[0]);
  y2[1]=-y2[1]-lamsq*(y0[0]-4*y1[0]+b*y2[0]-4*y3[0]+y4[0]);
         :
  y17[1]=-y17[1]-lamsq*(y15[0]-4*y16[0]+b*y17[0]-4*y18[0]+y19[0]);
  y18[1]=-y18[1]-lamsq*(y16[0]-4*y17[0]+c*y18[0]-2*y19[0]);
  y19[1]=-y19[1]-lamsq*(2*y17[0]-4*y18[0]+d*y19[0]);
}
```

Figure 6.11. Design of Algorithm 6

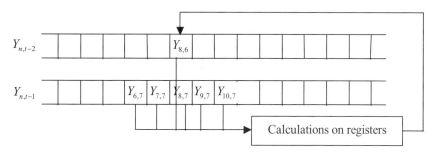

Figure 6.12. Re-calculating in two time steps

Algorithm 7: Nested Loops Two-element Array and Rotation
Algorithm 7, as shown in Figure 6.15, is based on improvements achieved with Algorithm 6. Additionally, the notion of nested loops was incorporated. The advantages and disadvantages of this approach were identified earlier and remain true for this particular algorithm.

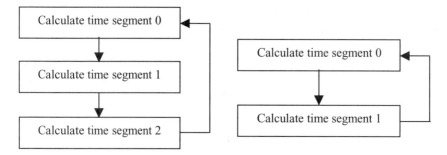

Figure 6.13. Description of Algorithm 2   **Figure 6.14.** Description of Algorithm 6

```
Loop {
    // Step 1
    y[0][0]=-y[0][0]-lamsq*(a*y[0][1]-4*y[1][1]+y[2][1]);
    y[1][0]=-y[1][0]-lamsq*(-4*y[0][1]+b*y[1][1]-4*y[2][1]+y[3][1]);
    for (i=2; i<18; i++)
      y[i][0]=-y[i][0]-lamsq*(y[i-2][1]-4*y[i-1][1]+b*y[i][1]-4*y[i+1][1]+y[i+2][1]);
    y[18][0]=-y[18][0]-lamsq*(y[16][1]-4*y[17][1]+c*y[18][1]-2*y[19][1]);
    y[19][0]=-y[19][0]-lamsq*(2*y[17][1]-4*y[18][1]+d*y[19][1]);
    // Step 2
    y[0][1]=-y[0][1]-lamsq*(a*y[0][0]-4*y[1][0]+y[2][0]);
    y[1][1]=-y[1][1]-lamsq*(-4*y[0][0]+b*y[1][0]-4*y[2][0]+y[3][0]);
    for (i=2; i<18; i++)
      y[i][1]=-y[i][1]-lamsq*(y[i-2][0]-4*y[i-1][0]+b*y[i][0]-4*y[i+1][0]+y[i+2][0]);
    y[18][1]=-y[18][1]-lamsq*(y[16][0]-4*y[17][0]+c*y[18][0]-2*y[19][0]);
    y[19][1]=-y[19][1]-lamsq*(2*y[17][0]-4*y[18][0]+d*y[19][0]);
}
```

Figure 6.15. Design of Algorithm 7

6.6.2 Experimentation and Results

Experimental Set-up

The PERL 5.6.0 program interpreter was used to generate a multitude of different C source code files for different algorithms, on the basis of number of beam segments. Figure 6.16 illustrates the general experimental set-up designed for automation of algorithm creation and evaluation. This enables significant numbers of algorithms to be rigorously tested in a time-efficient manner. Program parameters such as number of beam segments are examined on a very fine scale. The automatic source code generation eliminated the risk of bugs and coding errors.

The set-up consists of three phases: generation of source code, compilation of the source code and execution of the compiled code and collection of output data for further analysis. A meta-generation PERL script was employed to produce a shell script for each phase, which issues consecutively the commands required to perform the various tasks on a number of test samples.

The target environment for all experiments was based on a Linux distribution with 2.2.x kernel, which is a multi-user multitasking operating system. Hence, a homogeneous and continuous availability of system resources could not be guaranteed. The hardware platform was not shared among other users at the time of the experiments, but system services might have required resources at some instances. Finally it must be noted that the results and subsequent observations of the experiments depended on the type of target hardware used and might not be reproducible on different hardware architectures.

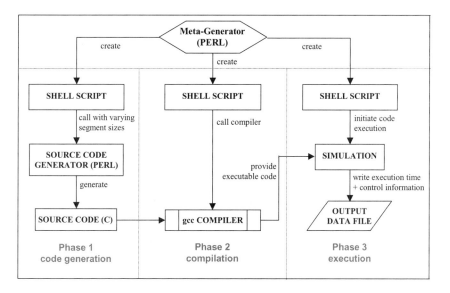

Figure 6.16. Experimental set-up using PERL script

Performance of the Algorithms
This section explores the relative performance of the seven algorithms outlined in the previous section. The first investigation into algorithm performance was performed with respect to memory access time with a set of four experiments. In this case, the number of beam segments was fixed while the number of iterations was increased. Figures 6.17, 6.18 and 6.19 depict the performance of Algorithm 1 and Algorithm 2 for 40, 60 and 80 beam segments, respectively.

It is noted that the execution time for 40 segments (as shown in Figure 6.17) with both algorithms increases linearly with the number of iterations. Algorithm–2 performs approximately 19.5% better than Algorithm 1, independent of the number of iterations used. Therefore, the design of Algorithm 2 appears to be superior to Algorithm 1 for 40 segments on the hardware considered.

Performances achieved in implementing Algorithm 1 and Algorithm 2 for 60 segments are similar, as shown in Figure 6.18. To explore the issue further, performance achieved in implementing the algorithms with 80 segments was considered. Figure 6.19 shows the execution times thus achieved. As compared with the performances achieved in implementing algorithm 1 and Algorithm 2 for 40 segments, it is noted that Algorithm 1 performed faster than Algorithm 2 with 80 segments. This is the reverse of the situation with 40 segments. For 80 segments Algorithm 1 outperformed Algorithm 2 in terms of execution time by approximately 38.8%.

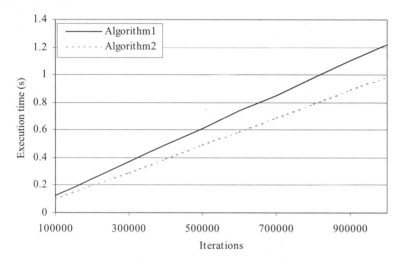

Figure 6.17. Execution time in implementing Algorithm 1 and Algorithm 2 for 40 segments

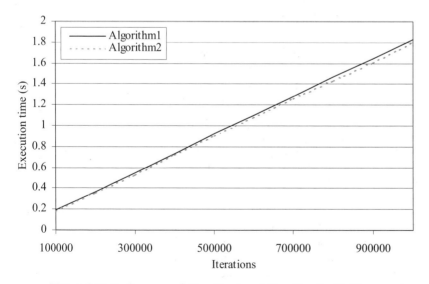

Figure 6.18. Performance of Algorithm 1 and Algorithm 2 with 60 segments

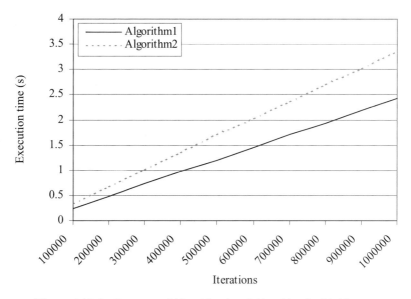

Figure 6.19. Performance of Algorithm 1 and Algorithm 2 with 80 segments

Finally, it is noted that Algorithm 2 offers better performance in respect of execution time with less than 60 segments. Algorithm 2 outperforms Algorithm 1 with a large number of segments (above 60).

To explore the trend in performance of the algorithms with number of beam segments further, a new set of experiments was conducted with a finer scale and larger range of segment numbers for Algorithm 1 and Algorithm 2. During these experiments the number of iterations for each simulation was fixed at 250,000. The results of these experiments confirm some of the conclusions made in the previous experiments. Figure 6.20 shows the performance of Algorithm 1 and Algorithm 2 for a fixed number of iterations and variable segments. It is noted that the performance of Algorithm 2 significantly decreases with a transition between segment sizes of 60 and 70. Similar performance degradation of Algorithm 1 is noted with a transition between segment sizes of 120 and 150. As a whole, Algorithm 2 shows better performance than Algorithm 1, except in the region of 70 to 130 segments.

Figure 6.21 shows the execution times achieved with Algorithm 1 and Algorithm 3. It is observed that the design mechanism of large arrays used in Algorithm 3 has significant weakness compared to Algorithm 1. Indeed it was noted that Algorithm 3 was outperformed by all other algorithms. This huge degradation could be due to large memory allocation (in turn, memory access) for Algorithm 3. Therefore, this approach is not considered in subsequent analysis.

As discussed earlier, the design of Algorithm 4 and Algorithm 5 originated from the mechanisms presented in Algorithm 1 and Algorithm 2, respectively. Moreover, these algorithms incorporate a fixed number of instructions in the source code related to an increase in the number of segments. Thus, the number of instructions within the main loop would not exceed the instruction portion of the system's cache. As a result a linear increase of program execution time in relation to the number of segments would be expected.

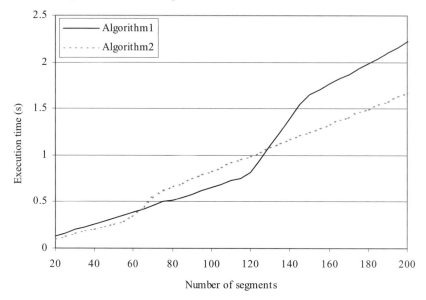

Figure 6.20. Execution time to implement Algorithm 1 and Algorithm 2 for 250,000 iterations

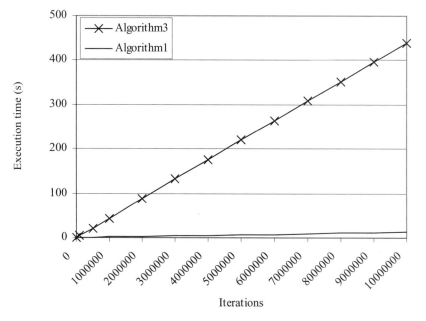

Figure 6.21. Execution time to implement Algorithm 1 and Algorithm 3 with 20 segments

Figure 6.22 shows the performance of Algorithm 1, Algorithm 2, Algorithm 4 and Algorithm 5. It is noted that Algorithm 1 and Algorithm 2 with their corresponding conventional approaches still outperformed the algorithms with nested loops (Algorithm 4 and Algorithm 5). However, with the new method in Algorithm 5 a linear characterisation with increases in the number of segments is achieved. This provides another important piece of evidence supporting the theory that instruction-level caching issues are responsible for the non-linear behaviour of Algorithm 1 and Algorithm 2 in relation to the number of beam segments.

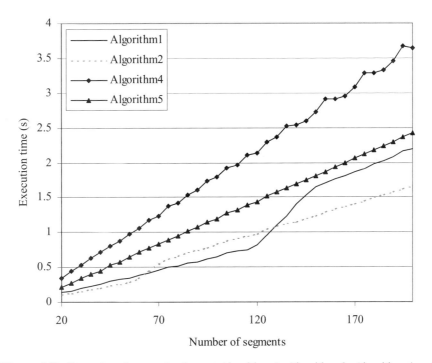

Figure 6.22. Execution times to implement Algorithm 1, Algorithm 2, Algorithm 4 and Algorithm 5 with 250,000 iterations

Considering the issue of the linear performance observed with Algorithm 4 and Algorithm 5, a different algorithm approach was pursued to reduce the number of instructions in the primary loop by means of algorithm design. Algorithm 6 was accordingly developed so as to reduce the number of instructions, compared to Algorithm 2. It is noted that the number of instructions in Algorithm 6 increased two-fold rather than three-fold.

Figure 6.23 depicts the execution times for Algorithm 1, Algorithm 2 and Algorithm 6. The transition in Algorithm 6 towards weaker performance occurred halfway between the transitions of Algorithm 1 and Algorithm 2. In spite of being outperformed by Algorithm 1 over the narrow band 90 to 125 segments, Algorithm 6 offers the best performance overall. Thus, the design mechanism employed in

Algorithm 6 can offer potential advantages in dynamic simulation, signal processing and control applications.

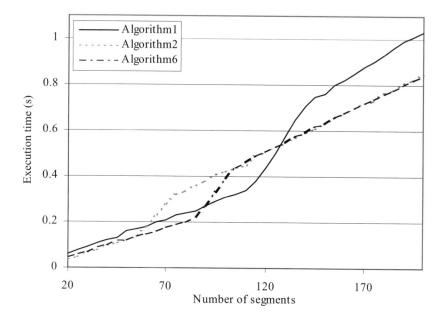

Figure 6.23. Performance comparison of Algorithm 1, Algorithm 2 and Algorithm 6

Figure 6.24 shows the execution time for all algorithms except Algorithm 3, which was discarded due to its low performance. It is noted that Algorithm 6 achieved the best performance except over a small region, while Algorithm 4 achieved the lowest performance among the algorithms. It is also noted that the performances of Algorithm 2 and Algorithm 6 are at a similar level, except over a small region where Algorithm 2 has a transition in performance. On the other hand, Algorithm 1 achieved the best performance over a small region, which is just after the transition of Algorithm 6 up to the transition of Algorithm 1 itself. Among the algorithms that achieved linear performance, Algorithm 7 shows the best performance. In addition, for higher numbers of segments, the performance of Algorithm 7 is even better than Algorithm 1.

6.7 Summary

This chapter has introduced and explored analysis and design of several signal processing and control algorithms, considered for real-time implementation. These include adaptive active vibration control, adaptive filtering, spectral analysis and simulation of a flexible manipulator system. The data and control dependencies, regular and irregular features of the algorithms, real-time computing methodology for uniprocessor and multiprocessor systems, including partitioning and task allocation of the algorithms for heterogeneous and homogeneous architectures have been discussed. This establishes the basis for suitable coding of the algorithms to exploit computer domains of sequential and parallel nature.

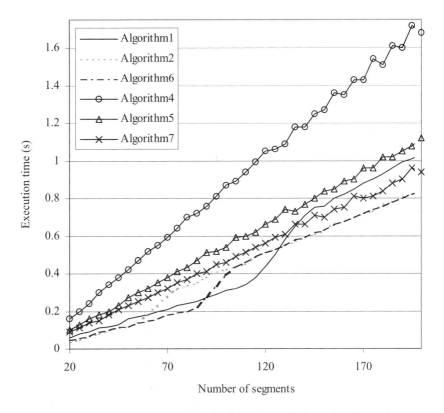

Figure 6.24. Execution times of the algorithms for a fixed number of iterations

6.8 Exercises

1. What is meant by the term 'algorithm'? Can one design a number of algorithms to solve a particular problem? If yes, give an example.

2. Indicate the differences between data dependence and control dependence. Which one has greater impact on performance of an application implemented in real time? Discuss the mechanisms to overcome the problems associated with data and control dependence.

3. Describe the granularity and regularity of an algorithm. Indicate the differences between the hardware and task granularity. What real-time computing issues can be considered to achieve higher task granularity and, in turn, potentially profitable parallelism?

4. Distinguish between theoretical computational complexity of Equations 6.1 and 6.30. Implement both equations using the same compiler on the same computing domain and describe the real-time performances.

5. Consider the algorithm

$$A(x,t) = A(x-1,t) + b^2 A(x-2,t) - c^2 A(x-3,t)$$

where, A is considered as an array, the initial value of the array can be considered zero, b and c are small constants, t represents the number of iterations and the value of x is 40. Design this algorithm based on the mechanisms of Algorithm 2 and Algorithm 6, as discussed in the case study, for 500000 iterations. Implement the algorithms on the same computing domain and compare the performances.

7. Microprocessors: The Processing Elements of Parallel Architectures

7.1 Objectives

- To introduce the microprocessor as the brain of sequential and parallel architectures.
- To introduce and discuss the classification of microprocessors.
- To provide a flavour of the evolution of microprocessors.
- To discuss features of various types of microprocessors.

7.2 Introduction

A microprocessor is considered to be the brain of a computer, possessing the main functional units that perform, control and monitor the activities of the computer. As such the characteristics and performance of a computer architecture, whether sequential or parallel, are defined by the microprocessor used to build it. The main functional units of a microprocessor include ALU, CUs and memory, referred to as registers. The ALU performs all arithmetical and logical operations as demanded by the program. For example, adding, subtracting, multiplying, dividing and comparing two numbers are performed by the ALU. The CU is the brain of the computer, and as such is the main functional unit controlling and synchronising all activities of the microprocessor. The CU performs tasks such as decoding, executing, storing and fetching data. Moreover, it synchronises overall activities of the microprocessor. The registers are used to hold data in various forms as operated on by the microprocessor. The main functional units of a microprocessor incorporate many subunits, each of which perform specific tasks. Other functional units associated with the activities of a microprocessor worth mentioning include floating-point unit (FPU), memory management unit (MMU), branch processing unit (BPU), vector processing unit (VPU), cache memory and built-in clock. The FPU deals with floating-point (non-integer) operations. Previously, the FPU was an external co-processor but nowadays it is integrated on-chip to speed up operations. The MMU translates addresses of an application into physical memory addresses.

This allows an operating system to map an application code and data in different virtual address spaces, which allows the MMU to offer memory protection services. The BPU predicts the outcome of a branch instruction, aiming to reduce disruptions in the flow of instructions and data into the processor when an execution thread jumps to a new memory location, typically as the outcome of a comparison operation or the end of a loop. The VPU handles vector-based, SIMD instructions that accelerate graphics operations. The level 1 (L1) cache memory of a microprocessor stores the most frequently used instructions and data. The microprocessor can access cache memory much faster than random access memory (RAM). Another important part of a microprocessor is its built-in clock. The clock determines the maximum speed at which other units can operate and helps synchronise related operations. Clock speed is measured in megahertz and, increasingly, gigahertz. The fastest commercial processors currently (November 2002) operate at about 3 GHz, or 3 billion clock cycles per second.

The characteristics of a microprocessor can be customised to a certain extent by manipulating the characteristics of its constituent functional units through appropriate design and adding extra units as required. As such, some microprocessors are well suited to general-purpose work, some are suited to DSP-based algorithms and some are suited to vector-based algorithms. There has been ongoing effort among scientists and engineers to enhance the performance of microprocessors and, accordingly, have developed various types to meet general and specific requirements. As a result, the processors in current computers have grown tremendously in performance, capabilities and complexity over the past decade. Clock speed has increased immensely, and size has reduced and the number of transistors packed on them has soared.

Performance enhancement of microprocessors over the years has boosted the performance of both sequential and parallel computing equally. As mentioned in Chapter 1, a parallel architecture comprises more than one microprocessor. Thus, the better the performance of a microprocessor, the better the performance of the corresponding parallel architecture. A parallel architecture could be designed incorporating microprocessors of the same type or different types and employed to execute algorithms according to their nature. To find an exact match between an algorithm and a parallel architecture is a matter of knowledge, experience and research but as algorithms of different nature exist so do microprocessors and parallel architectures. However, to design parallel architectures and to use them to execute algorithms as fast as possible one has to have sufficient knowledge of microprocessors of different kinds, their evolution and future trends. With this theme in mind, this chapter is designed to describe the characteristics of various types of microprocessor and the evolution of some of the significant microprocessor series that currently exist in the market.

7.3 Microprocessors: Classification

Various types of microprocessor exist in the market: in many cases their characteristics are very distinct; there are cases where their characteristics overlap. Thus, broad levels of classification are introduced in this chapter. This incorporates classifications based on the purpose (application) processors have been developed for and on the nature of their instruction set.

7.3.1 Classification Based on Application

Owing to the nature of target applications, microprocessors, in general, can be classified into two types, general-purpose and special-purpose.

General-purpose Microprocessors
As the name implies, general-purpose microprocessors are developed with the intention to use them as multipurpose devices in a wide range of applications. Today, a general-purpose microprocessor is not only used as the heart of the microcomputer but also in many other instruments in such applications as aerospace, control of structures, different kinds of industries and medical instruments. The Intel 4004 is regarded as the first general-purpose microprocessor to unveil the era of microprocessors. It was built by Intel in 1971 in response to the request of Busicom, a Japanese company, who wanted to use it to build a calculator with incredible overkill. Intel subsequently built the 8008 microprocessors for specific applications, like the 4004, and in 1974 built the 8080, which was used as the central processor of many of the early home computers. Since then many companies have developed varieties of general-purpose microprocessors some of which will be discussed in this chapter.

Special-purpose Microprocessors
Special-purpose processors are designed for use in specific applications or for meeting specific requirements that general-purpose processors cannot discharge. There are several types of special-purpose processors commonly used in various application areas. Among these, DSP devices, vector processors and pipeline processors are worth mentioning.

Digital signal processing devices are special types of processors developed to process digital signals. The signals that we experience in our day-to-day life (real world) such as sound, temperature, electrical voltage or current, pressure, the output of transducers, are analogue signals. These signals require to be processed through mathematical means to extract, analyse and transmit information with ease and hence they are represented first in the most suited language of mathematics, i.e., digital form (binary forms). DSP devices are specifically designed to process such signals so as to meet real-time requirements of an application. This real-time capability makes a DSP device perfect for applications that cannot endure delays. The DSP process is shown in Figure 7.1.

Some of the advantages of DSP devices compared to general-purpose processors are:

- Real-time performance.
- More flexibility.

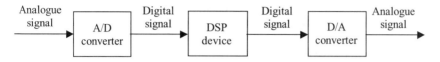

Figure 7.1. Digital signal processing

- More reliability.
- Increased system performance.
- More cost effectiveness.

Vector Processors
Vector processors are designed to execute arithmetic operations on elements of arrays, called vectors, very rapidly. These processors are especially useful in high-performance scientific computing dominated by vector-intensive algorithms, i.e., where vectors and matrices are quite common.

The architecture of a vector processor is designed so as to handle vector and vector-based algorithms efficiently. For example, some vector processors contain specially designed registers called vector registers. A vector register contains several elements of a vector at one time whereas a general-purpose or floating-point register holds a single value. A vector processor reads elements of a vector directly from the memory into the vector register. In these processors, all arithmetic or logical vector operations are register–register operations; i.e., they are performed only on vectors that are already in the vector registers and hence these processors are termed as vector–register vector processors. Another type of vector processor, called a memory–memory vector processor, allows the vector operands to be fetched directly from memory to different vector pipelines and the results to be written directly to memory.

Pipeline Processors
Pipelining is a method of decomposing a sequential process into suboperations, with each subprocess executed in a dedicated segment that operates concurrently with all other segments. Each segment performs partial processing dictated by the way the task is partitioned. The result obtained from the computation in each segment is transferred to the next segment in the pipeline. The final result is obtained after the data have passed through all the segments.

Pipeline processors are designed to process large volumes of data quickly and efficiently (e.g. digitised images) as required in many applications especially real-time. These processors contain a number of specialised, high-performance computational elements arranged one after the other in an assembly line style configuration and each computational element is dedicated to perform a specific task such as fetch, decode, execute instructions.

7.3.2 Classification Based on Nature of Instruction Set

In terms of the nature of instruction sets computers are classified into two groups, CISC and RISC. These were briefly introduced in Chapter 2, and are further described here.

CISC Processors
The CISC architecture endeavours to develop and execute a program with fewer instructions than its RISC counterpart. This is achieved by sacrificing the number of cycles per instruction, i.e., it supports/accommodates more cycles per instruction. Most CISC hardware architectures have several characteristics in common, including:

- CISC architecture possesses complex instruction-decoding logic driven by the requirement of a single instruction to support multiple addressing modes.
- The CISC architecture has a small number of general-purpose registers as a result of having instructions that can operate directly on memory and the limited amount of chip space not dedicated to instruction decoding and execution.
- Many CISC designs set aside special registers for the stack pointer, interrupt handling, and so on. This can simplify the hardware design to some extent, at the expense of a more complex instruction set.
- The CISC architecture possesses a condition code register, which is set as a side effect of most instructions. This register reflects whether the result of the last operation is less than, equal to, or greater than zero, and records if certain error conditions occur.

RISC Processors
As opposed to CISC architecture, RISC processors attempt to reduce the cycles per instruction at the cost of number of instructions per program. These RISC "reduced instructions" require less transistors and hardware space than the complex instructions, leaving more room for general-purpose registers. The instruction format of RISC processors is very simple and fixed in size. This enables faster decoding of instructions and efficient use of pipelining in the control unit of the processor. Some of the advantages of the RISC architecture, compared to CISC architecture, are as follows:

- With a limited number of instructions, the instruction decoder can be built with less complexity, and hence faster decoding of instructions can be achieved. Support for pipelining is another factor for increased RISC speed. Since a simplified instruction set allows for a pipelined, superscalar design, RISC processors often achieve 2 to 4 times the performance of CISC processors using comparable semiconductor technology and the same clock rates. Another advantage of the less complex decoder unit is that more silicon space is available for the designer that can be utilised to add more internal registers, which can in turn be used to improve the speed of execution of programs.

- Because of the simpler instruction set, RISC processors use less chip space as compared to its CISC counterpart for extra functions, such as MMUs or floating-point arithmetic units (FPAUs). Smaller chips allow a semiconductor manufacturer to place more parts on a single silicon wafer, which can lower the per-chip cost dramatically.
- A RISC processor can be designed more quickly than a CISC processor due to the simpler hardware requirement of the RISC architecture. As a result, technical developments and design of RISC architecture are expected to occur faster than the corresponding CISC designs, leading to greater leaps in performance between generations.

7.4 Microprocessor: Evolution and Features

There are several companies in the world developing and marketing microprocessors of various kinds. Since the development of the first Intel 4004 microprocessor there has been a continually increasing effort to develop microprocessors with better and better performance and varied characteristics. As a result current processors can address varied requirements with a high degree of efficiency. Evolution of some of the well-known microprocessors and their important features are briefly described in this section.

7.4.1 General-purpose Processors

General-purpose microprocessors enjoy widespread use, as they are capable of meeting multifarious demands. Production of this type of processor is much higher throughout the world than that of the special-purpose microprocessor. Many companies such as Intel, Motorola, Compaq, Apple, Toshiba and NEC are dedicated to the development and marketing of general-purpose microprocessors. It is not possible to cover all the processors developed by different companies and as such only a selected set of microprocessors are discussed here.

Intel Processors
From its inauguration as a microprocessor developing company with the 4004 in 1971, Intel has dominated the world microprocessor market. Today, Intel microprocessors are used worldwide as the brains of millions of computers. Evolution of Intel's microprocessors are discussed here covering some of the Intel's main microprocessors developed over the years.

(1971) 4004 Microprocessor. The world's first microprocessor developed by Intel, Intel-4004, paved the way for personal computers (PCs). The 4-bit Intel 4004 ran at a clock speed of 108 Hz and contained 2300 transistors. It processed data in 4 bits, but its instructions were 8 bits long. The 4004 addressed up to 1 KB of program memory and up to 4 KB of data memory (as separate entities).

(1972) 8008 Microprocessor. The 8008 developed in 1972 was twice as powerful as the 4004. It was designed using 3500 transistors. The width of its address bus was 8 bits and it possessed 16 KB of addressable memory. It ran at a clock speed of 200 Hz.

(1974) 8080 Microprocessor. Developed in 1974, the 8080 microprocessor became the brains of the first personal computer--the Altair. Its clock speed was 2 MHz and the width of the address bus was 8 bits. It possessed an addressable memory of 64 KB. It was designed using 6000 transistors.

(1977) 8085 Microprocessor. Intel released the 8-bit 8085 microprocessor in 1977. It was an improved version of the 8080 and could run 769,230 instructions per second. It was the last as well as the most successful 8-bit microprocessor Intel produced. The main improvements in 8085 over the 8080 include its internal clock generator, internal system controller and higher speed (Brey, 2003).

(1978-79) 8086-8088 Microprocessor. The 8086 and 8088 were developed in 1978 and 1979 respectively. The 8086 was available at 5 MHz, 8 MHz and 10 MHz and the 8088 was available at 5 MHz and 8 MHz clock speeds. Both processors possessed address buses 16 bits wide and addressable memory of 1 MB. Both were designed using 29000 transistors. In fact, the 8088 was identical to 8086 except for its 8-bit external buses.

(1982) 286 Microprocessor. Released in 1982, the 286, also known as the 80286, was a 16-bit processor. It was available at 6 MHz, 8 MHz, 10 MHz, and 12.5 MHz clock speeds. Its address bus was 16 bits wide and it possessed an addressable memory of 16 MB. Moreover, it possessed a virtual memory of 1 gigabyte (GB). Its performance was three to six times higher than that of the 8086. It was also the first Intel processor that could run all the software written for its predecessor. This software compatibility made it very popular among the Intel processor users. Within 6 years of it release, there were an estimated 15 million 286-based personal computers installed around the world.

(1985) Intel 386™ Microprocessor. The Intel 386™ was the first 32-bit microprocessor possessing a 32-bit data bus and 32-bit memory address. It featured 275,000 transistors--more than 100 times as many as the original 4004. It was capable of "multitasking," meaning it could run multiple programs at the same time. It possessed an addressable memory of 4 GB and a virtual memory of 64 terabytes (TB). It was available at 16 MHz, 20 MHz, 25 MHz and 33 MHz clock speeds. One of the important features of the 386 was that it possessed an MMU that allowed memory resources to be allocated and managed by the operating system. Previously memory management was completely handled by software. The 80386 included hardware circuitry for memory management and memory assignment that improved its performance and reduced software overhead. The 80386 could move from virtual mode to real mode without resetting the processor. The instruction sets of the 80386 were upward compatible with the earlier 8086, 8088 and 80286 processors.

(1989) Intel 486™ DX CPU Microprocessor. Intel 486™ was designed using 1 million transistors. Its address bus was 32 bits wide. It possessed an addressable memory of 4 GB and virtual memory of 64 TB. It was available at 25 MHz, 33 MHz, 50 MHz clock speeds. It incorporated the necessary co-processor circuitry on the same slice of silicon. The 80486 chip contained an 8 KB cache memory and an improved 80387 numeric co-processor. This internal co-processor nearly doubled the performance of the processor. The 486 pioneered one-micron design rules, that is, the finest details etched into chip measure one micron across. The 486 helped minimise the effect of this memory slowdown by incorporating its own high-speed memory cache. The cache in a 486 was organised as four-way set associative design which essentially splits up its 8K total size as four small 2K caches, an arrangement that further enhances its performance, particularly in multi-threaded applications (Tokhi *et al.*, 1999a). The Intel 486™ processor was the first to offer a built-in math co-processor, which speeds up computing because it offloads complex math functions from the central processor.

(1993) Intel Pentium® processor. Intel's Pentium was Intel's first 64-bit microprocessor. It was designed using 3.1 million transistors. Its addressable memory was 64 GB and virtual memory 64 TB . Its performance was five times better than that of the Intel 80486 DX processor. It was initially available at 60 MHz and 66 MHz clock speeds. The data bus transfer speed was either 60 MHz or 66 MHz depending on the version of the Pentium. Later versions of Pentium also included additional instructions, called multimedia extensions or MMX instructions (Shaheed, 2000). The Pentium® processor allowed computers to more easily incorporate "real-world" data such as speech, sound, handwriting and photographic images. One of the most striking features of the Pentium was its dual integer processor facility. The Pentium could execute two instructions simultaneously which were not dependent on each other, with the help of two independent integer processors, called superscalar technology. This allowed the Pentium to often execute two instructions per clock period. A jump prediction technology speeding the execution of programs containing loops also boosted the performance of Pentium. Like the 80486 the Pentium also employed an internal floating-point co-processor to handle floating-point data. All these features improved the speed of the Pentium tremendously.

(1995) Pentium® Pro processor. Released in 1995 the Pentium® Pro processor was designed to fuel 32-bit server and workstation applications, enabling fast computer-aided design, mechanical engineering and scientific computation. The powerful Pentium® Pro processor contained 5.5 million transistors. Like the Pentium, it is a 32-bit microprocessor. Its addressable memory was 64 GB and virtual memory was 64 TB. It was available at 150 MHz, 166 MHz, 180 MHz, and 200 MHz clock speeds. In addition to the internal 16 KB L1 cache (8 KB for instruction and 8 KB for data) Pentium Pro processor contained a 256 level 2 (L2) cache allowing the processor to execute up to three instructions simultaneously, as opposed to Pentium, which can execute two integer operations simultaneously (Brey, 2003).

(1997) Pentium® II processor. Intel Pentium II processor was designed using 7.5 million transistors and was released in 1997. Pentium® II processor incorporated Intel MMX™ technology, which was designed specifically to process video, audio and graphics data efficiently. It was introduced in an innovative single edge contact (SEC) cartridge that also incorporated a high-speed cache memory chip. With this chip, PC users can capture, edit and share digital photos with friends and family via the Internet; edit and add text, music or between-scene transitions to home movies; and, with a videophone, send video over standard phone lines and the Internet. Pentium II was designed as a small circuit board instead of an integrated circuit (IC) to incorporate L2 cash on the board. This on-board L2 cache operated at a speed of 133 MHz and could store 512 KB of information. Like Pentium Pro its addressable memory was 64 GB and virtual memory was 64 TB. It was available at 200 MHz, 233 MHz, 266 MHz and 300 MHz clock speeds.

(1999) Pentium® III processor. Released in 1999, Pentium® III processor incorporated 9.5 million transistors, and was introduced using 0.25-micron technology. The Intel Pentium III processor is available at speeds ranging from 450 MHz to 1.33 GHz. The various versions are available with either a 133 MHz or 100 MHz system bus. The processor possesses P6 (Pentium Pro) dynamic execution micro-architecture including multiple branch prediction, data flow analysis and speculative execution, Internet streaming SIMD extensions (SSE), consisting of 70 instructions that enable advanced imaging, 3D streaming audio and video, speech recognition and an enhanced Internet experience, Intel® MMX™ media enhancement technology. The processor supports the high-performance dual independent bus (DIB) architecture. Versions also include an advanced transfer cache (ATC) and advanced system buffering (ASB) to meet higher data bandwidth requirements.

The Intel Pentium III processor includes two separate 16 KB L1 caches: one for instruction and one for data. The L1 cache provides fast access to recently used data, increasing overall performance of the system. The ATC of Pentium III consists of micro architectural improvements to provide a higher data bandwidth interface between the L2 cache and the processor core. Some versions of the Intel Pentium III processor include a discrete, off-die L2 cache. This L2 cache consists of a 256 KB unified, non-blocking cache that improves performance over cache-on-motherboard solutions by reducing the average memory access time and by providing fast access to recently used instructions and data. Performance of the Pentium III processor has also been enhanced over cache-on-motherboard implementations through a dedicated 64-bit cache bus.

(2000) Pentium® 4 processor. The Intel® Pentium® 4 processor is Intel's most advanced, most powerful processor for desktop PCs and entry-level workstations. It is based on Intel® NetBurst™ micro-architecture, which possesses a number of innovative features including hyper-pipelined technology, 533 MHz or 400 MHz system bus, execution trace cache (ETC) and rapid execution engine (REE). This architecture also possesses a number of enhanced features such as ATC, advanced dynamic execution (ADE), enhanced floating-point and multimedia unit and SSE 2.

The Pentium® 4 Processor is designed with 42 million transistors and circuit lines of 0.18 microns. The Intel's first microprocessor, the 4004, ran at 108 KHz (108,000 Hz), compared to the Pentium® 4 processor's initial speed of 1.5 GHz (1.5 billion Hz). The Pentium 4 processors are available at various clock speeds from 1.60 to 2.8 GHz, and is designed for desktop PCs as well as for entry-level workstations. The processor is upward compatible with previous generation Intel architecture processors.

In addition to the 8-KB data cache, the Pentium 4 processor includes an ETC that stores up to 12-K decoded micro-ops in the order of program execution. Two ALUs on the Pentium 4 processor are clocked at twice the core processor frequency. This allows basic integer instructions such as add, subtract, logical AND, logical OR, etc. to execute in one-half the clock cycle. The Pentium 4 processor expands the floating-point registers to a full 128-bit and adds an additional register for data movement which improves performance on both floating-point and multimedia applications.

The Pentium 4 processor is designed to deliver performance across applications, which include Internet audio and streaming video, image processing, video content creation, speech, three-dimensional (3D) computer-aided design (CAD), games, multimedia, and multitasking user environments.

(2001) Itanium™ processor. The Itanium™ processor is the first in a family of 64-bit products from Intel. The processor is based on Intel's explicitly parallel instruction computing (EPIC) design technology. It possesses 64-bit address and high memory bandwidth with three levels of caches: L1 cache of 32 KB, L2 cache of 96 KB and level 3 (L3) cache of 2 MB/4 MB memory. The processor is available at 733 MHz and 800 MHz clock rates. Its system bus frequency is 266 MHz. The Intel® Itanium® processor offers scalability. The Itanium architecture currently includes high capability for a variety of targeted applications.

The Intel Itanium® 2 processor. This is the second in a family of 64-bit products that bring performance and volume economics of the Intel architecture to the most demanding high-performance, technical and business-critical computing applications. The Intel® Itanium® 2 processor is also based on EPIC architecture. It also incorporates enhanced machine check architecture (MCA) with extensive error correcting code (ECC). The processor runs at 1 GHz or 900 MHz and is available with 3 MB or 1.5 MB integrated on-die L3 cache. Its L1 cache is 32 KB and the L2 cache is 256 KB. Its system bus is 400 MHz with 6.4 GB/s bandwidth. This bus is 128 bits wide. The processor offers high-end reliability and scalability features for business-critical computing and powerful solutions for vast amounts of data and users, high volumes of transactions and complex calculations.

The Intel Itanium architecture goes beyond RISC and CISC approaches by employing EPIC, which pairs extensive processing resources with intelligent compilers that enable parallel execution explicit to the processor. The Intel Itanium 2 processor is designed to support very large scale systems, including those employing thousands of processors.

A summary of the evolution of features of Intel processors is shown in Table 7.1 (Intel, 2002).

Table 7.1. Evolution of Intel microprocessors

Type	Date introduced	Clock speeds	Bus width	Number of transistors	Addressable memory	Virtual memory
4004	15 Nov. 1971	108 Hz	4 bits	2 300 (10 microns)	640 bytes	–
8008	01 April 1972	200 Hz	8 bits	3 500 (10 microns)	16 KB	–
8080	01 April 1974	2 MHz	8 bits	6 000 (6 microns)	64 KB	–
8086	08 June 1978	5 – 10 MHz	16 bits	29 000 (3 microns)	1 MB	–
8088	01 June 1979	5 & 8 MHz	8 bits	29 000 (3 microns)	1 MB	–
80286	01 Feb. 1982	6 – 12.5 MHz	16 bits	134 000 (1.5 microns)	16 MB	1 GB
386^{TM} DX	17 Oct. 1985	16 – 33 MHz	32 bits	275 000 (1 micron)	4 GB	64 TB
386^{TM} SX	16 June 1988	16 – 33 MHz	16 bits	275 000 (1 micron)	16 GB	64 TB
486^{TM} DX	10 Apr. 1989	25 – 50 MHz	32 bits	1.2M (1–0.8 micron)	4 GB	64 TB
486^{TM} SX	22 Apr. 1991	16 – 33 MHz	32 bits	1.185M (1 micron)	4 GB	64 TB
Pentium®	22 Mar. 1993	60 & 66 MHz	64 bits	3.1M (0.8 micron)	4 GB	64 TB
Pentium® Pro	01 Nov. 1995	150– 200 MHz	64 bits	5.5M (0.35 micron)	64 GB	64 TB
Pentium® II	07 May 1997	200 – 300 MHz	64 bits	7.5M (0.35 micron)	64 GB	64 TB
Pentium® III	26 Feb. 1999	0.65 – 1 GHz	64 bits	9.5M (0.25 micron)	64 GB	64 TB
Pentium® 4	Nov. 2000	1.3 – 2.8 GHz	64 bits	42M (0.13 micron)	64 GB	64 TB

It is noted in Table 7.1 that the number of transistors has grown significantly as the Intel microprocessor technology has evolved over the years. This is shown in Table 7.2. Gordon Moore, Chairman Emeritus of Intel Corporation, made his famous observation in 1965, just four years after the first planar IC was discovered. Moore observed an exponential growth in the number of transistors per IC and predicted that this trend would continue. This is called "Moore's Law", shown in Figure 7.2, which still holds today.

Table 7.2. Number of transistors used in Intel microprocessors (Intel, 2002)

Microprocessor type	Year of development	Number of transistors
4004	1971	2,250
8008	1972	2,500
8080	1974	5,000
8086	1978	29,000
286	1982	120,000
386™ processor	1985	275,000
486™ DX processor	1989	1,180,000
Pentium® processor	1993	3,100,000
Pentium II processor	1997	7,500,000
Pentium III processor	1999	24,000,000
Pentium 4 processor	2000	42,000,000

Motorola Processors
Motorola entered into the microprocessor business by releasing its first microprocessor MC6800 in 1974. Since then Motorola has continued to develop microprocessors with innovative features and emerged as a major competitor of Intel. Some of the Motorola's processors are described below (Motorola, 2002).

MC6800 and MC6809. Motorola released the MC6800 microprocessor after the Intel 8080 in 1974. It was an 8-bit microprocessor designed with 4000 transistors, and possessed only 78 instructions. The MC6809, released in 1982, was also an 8-bit microprocessor. The MC6809 was a major advance over its predecessor, the MC6800. It possessed two 8-bit accumulators and could combine them into a single 16-bit register. It also featured two 16-bit index registers and two stack pointers, which allowed for some advanced addressing modes. The MC6809 was source compatible with the MC6800, even though the MC6800 had 78 instructions and the MC6809 had only 59. Some instructions were replaced by more general ones, which the assembler would translate, and some were even replaced by addressing modes. The MC6809 was originally produced in 1 MHz and 2 MHz versions, but faster versions were produced later. The MC6809 had an internal clock generator. The MC6809E needed an external clock generator.

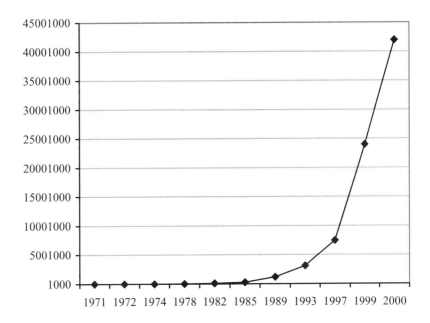

Figure 7.2. Moore's law; growth of number of transistors per IC

MC 68000. Motorola introduced its first 16-bit microprocessor, the MC68000, in 1979. It was capable of completing two million calculations per second and was used both to run and to write programs for scientific data processing and business applications. It was software compatible to a satisfactory extent. The processor had 8 general-purpose data registers, and 8 address registers, each 32 bits long. The data registers served as general-purpose accumulators and as counters. The MC68000 instructions were designed to comply with operands of three different lengths of 32 bits, 16 bits and 8 bits. A 32-bit operand corresponds to a long word, a 16-bit operand constitutes a word and an 8-bit operand corresponds to a byte (Hamacher *et al.*, 2002). The address registers hold information used in determining the address of memory operands. The address registers and address calculations involve 32 bits. However, in the MC68000, only the least significant 24 bits of an address are used externally to access the memory.

The memory of a MC68000 computer is organised in 16-bit words and is byte addressable. The Initial clock rate of the MC68000 was 8 MHz. Motorola ceased production of the MC68000 in 2000. A subtle problem with the MC68000 was that it could not easily run a virtual image of itself without simulating a large number of instructions. This problem persists in many modern versions of the architecture, which is rarely used in these applications.

MC68010. The Motorola MC68010 processor was released in the early 1980s. It resembled the MC68000 to a large extent. However, in addition to features of the MC68000 it possessed several instructions for register control. It also had the ability to save all of the processor state on an interrupt. This made it easier to use for virtual memory applications compared to the MC68000. The MC68010 was unable to gain popularity to the expected level. However, due to improved performance over the 68000, it enjoyed use in a number of Unix workstations, research machines, and in a number of Amiga machines in the 1980s.

MC68020. Motorola introduced the MC68020, the first true 32-bit microprocessor, in 1984. It contained 200,000 transistors on a three-eighths-inch square chip. It was much more powerful than its predecessors MC68000 and MC68010 because of notable enhancement in architectural features. In addition to its 32 bits internal and external data and address buses it possessed a 256-byte instruction buffer, arranged as 64 direct-mapped 4-byte entries. Although its data bus is 32-bit wide, the MC68020 can deal efficiently with devices that transfer 8, 16, or 32 bits at a time. The improvements of MC68020 over the MC68000 and MC68010 also include a 32-bit ALU and external data bus and address bus, and new instructions and addressing modes. The processor is dynamically adaptive to the data bus width requirements of a particular device in a manner that is transparent to the programmer. It can deal with devices of different data transfer sizes without knowing the actual size before a data transfer is initiated. The MC68020 possesses a proper three-stage pipeline. The new instructions included some minor improvements and extensions to the supervisor state, some support for high-level languages, and bit field manipulations. Although small, it made a significant difference in the performance of many applications. The MC68020 was used in many models of the Amiga and Apple Macintosh II series of personal computers and Sun 3 workstations.

MC68030. The MC68030 is a 32-bit microprocessor in Motorola's MC68000 family, with on-chip split instruction and data cache of 256 bytes each. The important additional features of the MC68030 over the MC68020 include another cache for data in addition to the instruction cache. Both caches are of the same size. The data cache organisation has 16 blocks of four long words each. The MC68030 also contains an on-chip MMU. The execution unit in the MC68030 generates virtual addresses (Hamacher et al., 2002). The cache access circuitry determines if the desired operand is in the cache, based on virtual addresses. The MMU translates the virtual address into a physical address in parallel with the cache access so that, in the case of a cache access miss, the physical address needed to access the operand in the main memory is immediately available. The MC68030 was used in many models of the Apple Macintosh II and Amiga series of personal computers as well as the Atari Falcon.

MC68040. The MC68040 is the first 680x0 family member with an on-chip FPU. It also contains MMU and is fully pipelined, with six stages. It possesses split instruction and data caches of 4 KB each. Instructions and data are transferred by two internal buses from the respective caches. The processor possesses monitoring circuitry to monitor activity on the external bus. As a result the MC68040 has been suitable for use in multiprocessor systems. Various versions of MC68040 were used in the Amiga, Apple Macintosh Quadra series of personal computers and in a number of workstations.

MC68060. The Motorola MC68060 is a 32-bit microprocessor and the latest member of the 680x0 family. It was introduced in the mid-1990s and was available with different clock rates ranging from 50 to 75 MHz. It is the highest performance processor of the Motorola 680x0 family with two to three times the performance capability of the MC68040. The MC68060 is characterised as a pipelined superscalar processor. Altogether there are six pipeline stages in the MC68060, of which four are basic stages and the other two are used for memory write back operation. Dynamic branch prediction is used to enhance smooth flow of instructions through pipeline. The main instruction processing hardware comprises three functional units: two integer units and a floating-point unit. The processor also possesses separate on-chip instruction and data catches. The MC68060 was used mainly in the Amiga machines and at a later stage production of the MC68060 was abandoned in favour of PowerPC chips. However, a number of chips, each with different sets of interfaces, were marketed under the name ColdFire and Dragonball.

PowerPC Processors
PowerPC is a RISC microprocessor architecture designed and created by the 1991 Apple-IBM-Motorola alliance, known as AIM. At the time several companies including IBM, Apple and Motorola recognised the demand for RISC-based machines in the market. IBM already had the performance optimisation with enhanced RISC (POWER) architecture, which had been introduced in the RS/6000, but a single-chip version was favoured. The alliance was made to develop RISC architecture based PowerPC processors and PowerPC based machines. They started their work based on the basic POWER1 and the PowerPC specifications. As a result PowerPC 601, the first single-chip implementation of the design, was released in 1992. PowerPC processors are used in Apple Macintosh, IBM RS/6000 computer, Amiga and in many other systems. The characteristics and features of some of the prominent PowerPC processors are described below.

PowerPC 601. As mentioned, PowerPC 601, a 32-bit RISC processor with 2.8 million transistors and 32 KB of on-chip cache was released in 1992. It is a highly integrated single-chip processor that combines a powerful RISC architecture, superscalar machine organisation, and a versatile high-performance bus interface. Various units comprising the PowerPC include instruction queue and dispatch unit, instruction fetch unit, branch processing unit, fixed-point execution unit, floating-point execution unit, memory management unit, cache, memory queue, bus interface unit, sequencer unit and common on-chip processor unit. The bus interface configurations provide a wide range of system bus interfaces, including pipelined, non-pipelined, and split transactions. The 601 pipeline structure has been optimised for high performance and concurrent instruction processing in each of the execution units. The fixed-point pipeline performs all integer ALU operations and all processor load and store instructions, including floating-point loads and stores. The FPU pipeline consists of four stages. The decode stage contains 32 double-word registers, the instruction decode logic, and the main pipeline control for the FPU. The branch instruction pipeline has only two stages. The first stage can dispatch, decode, evaluate, and, if necessary, predict the direction of a branch instruction in one cycle. On the next cycle, the resulting fetch can access new instructions from the cache. The floating-point instruction pipeline contains six stages and has been optimised for fully pipelined execution of single-precision operations.

The MMU of the 601 performs virtual-to-real address translation for load and store instructions. It also acts as a backup for instruction fetch address translations. The MMU provides support for segment-oriented, page-oriented, and block-oriented translations. The interface defines a 32-bit address bus and a 64-bit data bus. These buses are logically and physically decoupled from one another so that the protocol can support system bus organizations that use pipelined, non-pipelined, or even split bus transactions. The 601 was initially available at 50 MHz and 66 MHz clock speeds.

PowerPC 603. The PowerPC 603 microprocessor is the second member of the PowerPC family, and was designed using 1.6 million transistors. The 7.4 mm by 11.5 mm CMOS chip features on-chip 8 KB instruction and data caches coupled to a high performance 32/64-bit system bus. The 603 fixed-point unit (FXU) executes most integer instructions in a single cycle. The FPU is a pipelined unit with single-precision and double-precision instructions. The FXU contains general-purpose registers (GPRs); the FPU contains floating-point registers (FPRs). The system unit executes condition-register and other system-register instructions. The execution units write the result of a finished instruction to the proper rename bus and to the associated rename register.

The split 8 KB data and instruction caches have associated MMUs, which implement the PowerPC virtual environment architecture (VEA). The 603 pipeline possesses four major pipeline stages: fetch, dispatch, execute, and write back. The floating-point unit performs single-precision and double-precision floating-point operations compliant with the IEEE-754 floating-point standard. A three-stage pipelined arithmetic engine with multiply, carry-propagate-add, and write-back stages is used. A double-precision floating-point multiply-add instruction has a four-cycle latency with a two-cycle throughput. The bus interface unit (BIU) accepts bus requests from the instruction and data caches, and places the requests on the 603 external bus. The BIU includes address register queues, priority logic, and bus control state machines.

PowerPC 604 and 604e. The PowerPC 604, a successor to PowerPC 603, was announced in 1996. PowerPC 604e is an improved version of the PowerPC 604. Both processors are superscalar RISC processors from IBM Microelectronics and

Motorola. The PowerPC 604 can dispatch and complete up to four instructions in a single clock cycle. The processor operates on 32-bit instructions and integer data and on 64-bit double-precision or 32-bit single-precision floating-point data. The PowerPC 604 architecture consists of a program control unit, two simple integer ALUs, one complex integer unit, an FPU, a load/store unit, a branch unit and instruction and data caches. The PowerPC 604e possesses one additional functional unit called the CRU. This unit performs logical operations on the condition register. The PowerPC 604 has a single 32-bit address space. Separate 16 KB instruction and data caches are available on-chip. Twice as much cache RAM is available on the PowerPC 604e, which provides separate 32 KB instruction and data caches. Accesses to the instruction cache are 128 bits wide, providing the processor with four 32-bit instructions in a single cycle. Accesses to the data cache are 64 bits wide. However, since general-purpose registers on the PowerPC 604 are only 32 bits wide, the processor can only take advantage of the full 64-bit data cache access width when fetching or storing 64-bit double-precision floating-point variables. The processors are targeted at general-purpose desktop computing and have found their way into the Apple Macintosh line of personal computers and in Macintosh clones. The fastest version of the PowerPC 604 operates at a clock speed of 180 MHz. The PowerPC 604e can operate at a clock speed of 225 MHz.

PowerPC 620. PowerPC 620 is the first 64-bit implementation of the PowerPC architecture in a processor designed by using 7 million transistors. Key design features include 64-bit internal data paths, 64 KB of on-chip cache, six independent execution units, and a high-speed bus interface. The 620 is upward compatible with earlier PowerPC processors and can execute existing 32-bit PowerPC programs, as well as new 64-bit programs written specifically to exploit features on the 620. Various features of the 620 include, 133 MHz processor clock, 128 bit data bus, 40-bit address bus, six execution units, split 32 KB caches, built-in L2 cache interface. Another important feature of the 620 is the presence of two 32 KB on-chip caches. The 620, like the 603 and 604, implements a Harvard architecture with separate code and data paths. One cache handles the code path, and the other handles the data. Each cache has its own MMU and functions independently of the other.

The 620 possesses a high-performance branch-prediction mechanism that requires prediction logic plus 64-bit rename buffers and reservation stations to store speculated branch results. The data bus of the 620 is 128 bits wide. The bus interface of PowerPC 620 includes integral support for a unified (i.e., both code and data) L2 cache, whose size is configurable from 1 to 128 MB. The cache–interface signals can run at the same speed as the processor clock rates or at one half or one quarter their speed, which allows the construction of an L2 cache from slower RAM. This on-chip cache interface eliminates the extra clock cycles normally required to drive any external L2 cache logic.

MPC7450. Motorola announced the MPC7450 PowerPC microprocessor in 2001. The MPC7450 processor belongs to the Motorola's fourth-generation (G4) PowerPC family. The MPC7450 microprocessor has an advanced, deeper, seven-stage pipeline with two additional execution units. The L2 cache has been integrated onto the die for greater speed, and a 256-bit datapath to the L1 cache has been implemented. The MPC7450 processor also supports a large L3 cache with a 64-bit data path, and provides multiple SRAM options. Additionally, the MPC7450 microprocessor implements Motorola's high-bandwidth system bus, capable of achieving bandwidth performance of 1064 MB/s. which can provide up to five times the bus performance over third-generation processors.

HP's Alpha Processors
Digital Equipment Corporation (DEC) introduced Alpha processors in 1992. Compaq purchased DEC in 1998. However, Compaq and HP merged in 2002 and emerged with the name HP. Thus, the Alpha processors now belong to HP. The Alpha architecture is RISC architecture designed with 64-bit address and data sizes. The family possesses three processors namely 21064, 21164, and 21264. These processors are briefly described below.

Alpha 21064. The Alpha 21064 was designed using 0.75-micron CMOS technology. It contained 1.7 million transistors and was available at 200 MHz processing speed (Hamacher *et al.*, 2002). The Alpha 21064 is equipped with four independent processing units: an integer unit, a floating-point unit, a branching unit, and a memory load/store unit. The 21064 possesses a 32-entry integer register file with four read ports and two write ports. Each set of two read and one write ports is assigned to one pipeline. The instruction issue rate is two instruction per clock cycle. The FPR file also possesses 32 entries with three read ports and two write ports. The program counter of the 21064 is 64 bits wide. There are 8 KB data and instruction caches with optional off-chip write-back backup cache. The L1 cache has a block width of 32 bytes.

There are seven integer pipeline stages and 10 floating-point pipeline stages in 21064. The first four stages of both pipelines perform the same tasks: instruction fetch, branch prediction, instruction issue and decode and instruction issue and register read. The 21064 is equipped with a 2K-by-1bit branch history table with approximately 80% accuracy for dynamic branch prediction.

Alpha 21164. The Alpha 21164 was designed using 0.5-micron CMOS technology and was released in 1994. It contains 9.3 million transistors. Its fastest speed is 300 MHz. The 21164 possesses a 40-entry integer register file with four read ports and two write ports. 32 of the registers are used for integer operations and the eight are used as shadow registers in PALcode. There are nine ports in a floating-point register file: four read and two write ports for two pipelines, one read port for a floating-point store, and two write ports to allow two floating-point loads in one cycle.

The Alpha 21164 possesses 8-KB direct-mapped caches in L1 to provide a fast access time while its 96-KB L2 unified cache is on-chip and uses a three-way set associative model. The data blocks are 32 bytes in L1 cache and 64 bytes in L2 caches.

Alpha 21164 possesses seven stages of integer pipeline, nine stages of floating-point pipeline, and up to 12 stages of on-chip memory instruction operations. There are two integer pipelines and two floating-point pipelines.

Alpha 21264. The Alpha 21264, the latest processor of the series, is designed with 0.35-micron CMOS technology using 15 million transistors. It was released in 1998. The 21264 processor is equipped with an 80-entry integer register file and 72-entry FPR file. The processor has L1 and L2 caches of 64 KB and 1 MB to 16 MB, respectively. The L2 cache is direct-mapped and is placed off-chip. The block width of the L1 cache is 64 bytes. The data cache operates at twice the clock speed of the processor.

There are seven pipeline stages in the Alpha 21264 with an extra stage dedicated to register renaming. The first stage is an instruction fetch, which fetches four instructions in each cycle and predicts the target address of the next instruction in the instruction cache. Although the Alpha 21264 can fetch only four instructions in a single cycle, up to six instructions per cycle can be dynamically issued in the same cycle, as it possesses four integer and two floating-point pipelines. There are 31 integer and 31 floating-point registers that are visible to users. The 21264 possesses local and global branch predictors and is facilitated with a tournament branch prediction scheme to dynamically choose the result from one of these two.

The SPARC Architecture of Sun Microsystems
The SPARC architecture developed by SUN Microsystems is characterised as a RISC processor (Sun, 1994). It is considered as scalable architecture. SPARC processors are designed for use in high-performance workstations and servers. They are compatible with shared memory based multiprocessor systems. The first version of the SPARC architecture could handle 32-bit addresses and data. In contrast, recent versions of the architecture can handle both addresses and data as 64-bit values, and are implemented by the UltraSPARC series of processors. The UltraSPARC processors are backward compatible meaning that they can execute machine code from earlier versiosn of the architecture (Hamacher *et al.*, 2002).

The UltraSPARC I was manufactured using 0.5-micron CMOS technology and used a nine-stage pipeline organisation. It operates at a clock frequency of 167 MHz. The UltraSPARC II was manufactured with 0.25-micron CMOS technology using 5.4 million transistors. It possesses on-chip instruction and on-chip data caches of 16 KB. The UltraSPARC II processor is binary compatible with all versions of SPARC® processors. It is equipped with two integer execution units and two floating-point execution units.The most recent member of the family, the UltraSPARC III, uses a 14-stage pipeline. It incorporates four integer execution units and three floating-point units. The UltraSPARC III is manufactured in 0.18-micron technology. Clock frequencies of the versions range from hundreds of MHz to several GHz. The on-chip L1 data cache of the processor is 64 KB and the instruction cache is 32 KB. They are both four-way set associative, and operate with 32 byte blocks. The external L2 cache can be configured for 4 MB or 8 MB capacity. The UltraSPARC III provides extensive support for use in multiprocessor configurations accommodating as many as hundreds of processors.

The ARM Family

Advanced RISC Machines (ARM) Limited has developed a number of families of processors. These include ARM7, ARM9, ARM10E and ARM SecurCore (ARM, 2002).

The ARM7 family. The ARM7 family is a range of low-power 32-bit RISC microprocessor cores, which include the ARM7TDMI, ARM7TDMI-S and ARM7EJ-S processor cores and the ARM720T cached processor. The ARM7TDMI and the ARM7TDMI-S core enable system designers to build embedded devices requiring small size, low power and high performance. The ARM7TDMI operates at 88 MHz–133 MHz. The ARM7EJ-S core provides the flexibility necessary to build Java-enabled, real-time embedded devices. The ARM720T is designed for devices using a platform operating system, such as Windows CE and Symbian OS. In addition to the CPU core, the ARM720T features an 8 KB unified cache and MMU. It operates at 75 MHz–100 MHz clock speeds.

The ARM9 family. The ARM9 family is built around the ARM9TDMI processor core and incorporates the 16-bit Thumb instruction set. The ARM9 Thumb family includes the ARM920T, ARM922T and ARM940T cached processor macrocells. The ARM920T and ARM922T are based around the ARM9TDMI 32-bit RISC CPU and feature instruction and data caches, and MMU enabling support for all major operating systems. The ARM922T is identical to the ARM920T but has 8K/8K caches while the ARM920T has 16K/16K caches. The ARM940T hard macrocell is suitable for a wide range of embedded control applications. The ARM940T is based on the ARM9TDMI 32-bit RISC CPU and features 4K/4K instruction and data caches, and a memory protection unit (MPU) enabling support for many real-time operating systems.

The ARM9E family of products are DSP-enhanced 32-bit RISC processors, suited for applications requiring a mix of DSP and microcontroller performance. The family includes the ARM926EJ-S, ARM946E-S and ARM966E-S processor macrocells. The ARM966E-S core possesses separate directly connected instruction and data tightly coupled memory, which have flexible sizes and run at the processor clock speed. The ARM966E-S supports ARM's real-time trace technology with the optional ETM9 embedded trace macrocell. The ARM966E-S features a simple memory map providing an area and power efficient solution for applications, which do not require complex memory management support.

ARM10E Thumb® family. The family features 64-bit load-store micro-architecture, IEEE754 compatible floating-point co-processor with vector operations, easy system integration, and a complete set of hardware and software development tools. The ARM10E family consists of the ARM1020E and ARM1022E processor cores which have 32K/32K and 16K/16K caches respectively. The ARM1020E and ARM1022E macrocells incorporate the advanced ARM10E™ integer core, instruction and data caches, full-features memory management unit. The ARM1022E is identical to the ARM1020E macrocell but has 16K/16K caches while the ARM1020E has 32K/32K caches.

The ARM1026EJ-S macrocell is a fully synthesisable processor. A Jazelle technology enhanced 32-bit RISC ARM10EJ-S CPU with extensive 64-bit internal bussing is combined with configurable instruction and data caches, configurable tightly coupled memories, support for parity protection on SRAM arrays, memory management and protection units, vector interrupt controller interface and advanced vector floating-point support.

ARM SecurCore family. ARM's SecurCore family incorporates special security features to help resist power and timing attacks. The ARM SecurCore family includes the SC100, SC110, SC200 and SC210 cores based on low-power 32-bit RISC technology. SC100 implements the key SecurCore features in a small 35K gate implementation. The SecurCore SC110 is identical to the SC100 but incorporates a cryptographic co-processor to accelerate execution of standards-based cryptographic algorithms. The SecurCore SC200 core and the SC210 macrocell incorporate Jazelle technology acceleration for Java card applications. The SecurCore SC210 is built on the SC200 by incorporating a cryptographic co-processor to accelerate execution of standards-based cryptographic algorithms.

7.4.2 Special-purpose Processors

The DSP device and vector processor are two types of special-purpose processors used extensively in various applications. The evolution of these processors is considered here.

DSP Devices
Digital signal processing devices are in great demand due to their special features and wide applicability in various fields. As such, many companies produce DSP devices. One of the key players in the field of DSPs is Texas Instruments. Its TMS320CXX series is well known. Some of the processors of this series are discussed here (Texas Instruments, 2002).

TMS320C1X series. Texas Instruments have delivered several processors in this series, which include the TMS320C10, TMS320C14, TMS320C16 and, TMS320C17. These processors possess the following common features:

- Performance up to 8.77 million instructions per second (MIPS).
- Object code compatible.
- 144/256-word on-chip data RAM.
- 1.5K/4K/8K-word on-chip program ROM.
- 4K-word on-chip program EPROM.
- 32-bit ALU/accumulator.
- Eight input/output channels.
- Dual-channel serial port.
- Simple memory and I/O interface.

The C10 provides the core CPU used in all other 'C1x devices. Its microprocessor operates at 5 MIPS. It provides a parallel I/O of 8 × 16 bits. Three versions with cycle times of 160, 200, and 280 ns are available. TMS320C14 devices, using the 'C10 core CPU, offer expanded on-chip RAM, and ROM or EPROM ('E14/P14), 16 pins of bit selectable parallel I/O, and I/O mapped asynchronous serial port, four 16-bit timers, and external/internal interrupts. The 'C14 devices can provide for microcomputer/microprocessor operating modes. Three versions with cycle times of 160 ns are available. TMS320C15 devices are also a version of the 'C10, offering expanded on-chip RAM, and ROM or EPROM ('E15/P15). The 'P15 is a one-time programmable (OTP), windowless EPROM version. These devices can operate in the microcomputer or microprocessor modes. Five versions are available with cycle times of 160 to 200 ns. TMS320C16 offers on-chip RAM of 256 words, an expanded program memory of 65K words, and a fast instruction cycle time of 114 ns (8.77 MIPS). TMS320C17 consists of five major functional units: the 'C15 microcomputer, a system control register, a full-duplex dual channel serial port, and a co-processor port. The dual-channel serial port is capable of full-duplex serial communication. The co-processor port allows the 'C17 to act as a slave microcomputer or as a master to a peripheral microcomputer.

TMS320C20X series. This series has only one processor: TMS320C206. It is fabricated with static CMOS IC technology. The architectural design is based upon that of the TMS320C20x series and is optimised for low-power operation. The combination of advanced Harvard architecture, on-chip peripherals, on-chip memory, and a highly specialised instruction set is the basis of the operational flexibility and speed of the 'C206. The 'C206 offers the following advantages over TMS320C1X Series:

- Enhanced TMS320 architectural design for increased performance and versatility.
- Advanced IC processing technology for increased performance.
- 'C206 devices are pin- and code-compatible with 'C203 and 'F206 devices.
- Source code for the 'C206 DSP devices is software-compatible with the 'C1x and 'C2x DSP devices.
- New static-design techniques for minimising power consumption and increasing radiation tolerance.

TMS320C3X series. The TMS320C3x's internal bus and special DSP instruction set have the speed and flexibility to execute up to 80 million floating-point operations per second (MFLOPS). The TMS320C3x optimises speed by implementing functions in hardware that other processors implement through software or microcode. This hardware-intensive approach provides performance previously unavailable on a single chip. The TMS320C3x can perform parallel multiply and ALU operations on integer or floating-point data in a single cycle. Each processor possesses a general-purpose register file, a program cache, internal dual-access memories, one direct memory access (DMA) channel supporting concurrent I/O, and a short machine-cycle time. General-purpose applications are greatly enhanced by the large address space, multiprocessor interface, internally and externally generated wait states, one external interface port, two timers, one serial port, and multiple-interrupt structure. High-level-language support implemented through a register-based architecture, large address space, powerful addressing modes, flexible instruction set, and floating-point arithmetic. The TMS320C30 is a 32-bit floating-point processor manufactured in 0.7-mm triple-level-metal CMOS technology. The TMS320C31 is 32-bit, floating-point processor manufactured in 0.6 μm triple-level-metal CMOS technology.

TMS320C4X series. TMS320C40 and TMS320C44 DSP devices belong to this series. The '320C40 DSP devices are 32-bit, floating-point processors manufactured in 0.72 μm, double-level metal CMOS technology. The '320C40 is a part of the fourth generation DSP devices from Texas Instruments and is designed primarily for parallel processing. This processor has two versions: TMS320C40-50, TMS320C40-60 with 50 MHz and 60 MHz clock speeds respectively. They can execute 25 MIPS and 30 MIPS respectively and possess cycle times of 40 ns and 30 ns respectively. Both versions have 2K of RAM, two timers, six (12%) DMA and 4G×32 parallel ports. They also possess six communications ports, six-channel DMA co-processor, single-cycle conversion to and from IEEE-754 floating-point format, source-code compatible with TMS320C3x, single-cycle 40-bit floating-point, 32-bit integer multipliers, twelve 40-bit registers, eight auxiliary registers and 14 control registers.

The TMS320C44 is also part of the TMS320C4x generation of DSP devices as well. It is a 32-bit, floating-point processor manufactured in 0.72 μm double-level-metal CMOS technology. In addition to the features contained in the 320C40, the processor possesses two timers, two identical external data and address buses supporting shared memory systems, 128 MB program/data/peripheral address space, separate internal program-, data-, and DMA-co-processor buses for support of massive concurrent I/O of program and data.

TMS320C5X series. The TMS320C5x generation of the Texas Instruments TMS320 DSP devices is fabricated with static CMOS IC technology. The architectural design of the TMS320C5x is based upon the TMS320C25. The combination of advanced Harvard architecture, on-chip peripherals, on-chip memory, and a highly specialised instruction set is the basis of the operational flexibility and speed of the 'C5x devices. They execute up to 50 MIPS and offer

- Enhanced TMS320 architectural design for increased performance and versatility.
- Modular architectural design for fast development of spin-off devices.
- Advanced integrated-circuit processing technology for increased performance.
- Upward-compatible source code.
- Enhanced TMS320 instruction set for faster algorithms and for optimised high-level language operation.

- New static-design techniques for minimising power consumption and maximising radiation tolerance.

TMS32062X fixed-point DSP devices. The TMS320C6202, TMS320C6202B and TMS320C6203B devices are part of the TMS320C62x™ fixed-point generation of DSP devices in the TMS320C6000™ DSP platform. The C62x™ DSP devices are based on the high-performance advanced VelociTI™ very-long-instruction-word (VLIW) architecture developed by Texas Instruments. These DSP devices are suited to multichannel and multifunction applications.

The TMS320C6202/02B/03B has a performance capability of up to 2400 MIPS at 300 MHz. The C6202/02B/03B DSP devices possess the operational flexibility of high-speed controllers and the numerical capability of array processors. These processors have 32 general-purpose registers of 32-bit word length and eight highly independent functional units. The eight functional units provide six ALUs for a high degree of parallelism and two 16-bit multipliers for a 32-bit result. The C6202/02/03B can produce two multiply-accumulates (MACs) per cycle. This gives a total of 600 million MACs per second (MMACS) for the C6202/02B/03B device. The C6202/02B/03B also have application-specific hardware logic, on-chip memory, and additional on-chip peripherals.

The program memory of C6202/02B devices consists of two blocks, with a 128KB block configured as memory-mapped program space, and the other 128 KB block user-configurable as cache or memory-mapped program space. Data memory for the C6202/02B consists of two 64 KB blocks of RAM.

The C6203B device program memory consists of two blocks, with a 256 KB block configured as memory-mapped program space, and the other 128 KB block user-configurable as cache or memory-mapped program space. Data memory for the C6203B consists of two 256 KB blocks of RAM.

The TMS320C64x™ fixed-point DSP devices. This series comprise TMS320C6411, TMS320C6414, TMS320C6415 and TMS320C6416 processors. The TMS320C64x™ DSP devices are the highest-performance fixed-point DSP generation in the TMS320C6000™ DSP platform. Like TMS32062X Series, the TMS320C64x™ DSPs are based on the second-generation high-performance, advanced VelociTI™ VLIW architecture (VelocTI.2™). These DSPs are also suited to multichannel and multifunction applications. The C64x™ is a code-compatible member of the C6000™ DSP platform.

The C64x™ DSP device possesses the operational flexibility of high-speed controllers and the numerical capability of array processors. The C64x™ DSP core processor has 64 general-purpose registers of 32-bit word length and eight highly independent functional units—two multipliers for a 32-bit result and six ALUs—with VelociTI.2™ extensions. The VelociTI.2™ extensions in the eight functional units include new instructions to accelerate the performance in key applications and extend the parallelism of the VelociTI™ architecture. The C64x™ DSP also has application-specific hardware logic, on-chip memory, and additional on-chip peripherals similar to the other C6000™ DSP platform devices.

The C64x™ DSP device uses a two-level cache-based architecture and has a powerful and diverse set of peripherals. The L1 program cache is a 128-kilo bits (Kb) direct mapped cache and the L1 data cache is a 128-Kb two-way set-associative cache. The L2 memory/cache of the C6411consists of a 2 mega bits (Mb) memory space that is shared between program and data space. On the other hand, The L2 memory/cache of the C6414/C6415 and C6416 consists of an 8 Mb memory space that is shared between program and data space. L2 memory can be configured as mapped memory, cache, or combinations of the two.

The TMS320C64x™ DSP devices have a set of development tools which includes: a new C compiler, an assembly optimiser to simplify programming and scheduling, and a Windows™ debugger interface for visibility into source code execution.

The C6411 device possesses performance of up to 2400 MIPS at a clock rate of 300 MHz, whereas the C6414, the C6415 and the C6416 devices possess performance of up to 4800 MIPS at a clock rate of 600 MHz. The C6411 can produce two 32-bit MACs per cycle for a total of 600 MMACS, or eight 8-bit MACs per cycle for a total of 2400 MMACS. One the other hand, each C6414, the C6415 and the C6416 can produce two 32-bit MACs per cycle for a total of 1200 MMACS, or eight 8-bit MACs per cycle for a total of 4800 MMACS

The TMS320C67x floating-point DSP devices. TMS320C6701, TMS320C6711and TMS320C6711B are three processors belong to this series. The TMS320C67x™ DSPs compose the floating-point DSP family in the TMS320C6000™ DSP platform. These devices are also based on the high-performance, advanced VelociTI™ VLIW architecture. These are also well suited to multichannel and multifunction applications.

The TMS320C67x™ DSP devices possess the operational flexibility of high-speed controllers and the numerical capability of array processors. This processor has 32 general-purpose registers of 32-bit word length and eight highly independent functional units. The eight functional units provide four floating-/fixed-point ALUs, two fixed-point ALUs, and two floating-/fixed-point multipliers. The development tools of the TMS320C67X™ DSP devices include: a C compiler, an assembly optimiser, and a Windows™ debugger interface for visibility into source code execution.

The device performs up to 1G floating-point operations per second (GFLOPS) at a clock rate of 167 MHz. On the other hand, the C6711/C6711B can execute up to 900 MFLOPS at a clock rate of 150 MHz. The 'C6701 can produce two MACs per cycle for a total of 334 MMACS. The C6711/C6711B can produce two MACs per cycle for a total of 300 MMACS.

TMS320C8X, multimedia video processor. TMS320C80 and TMS320C82 belong to this series. The TMS320C80 is a single chip, MIMD parallel processor capable of performing over two billion operations per second. The TMS320C82 is also a single chip, MIMD parallel processor. However, it is capable of performing over 1.5 billion operations per second. Both processors consist of a 32-bit RISC master processor with a 120-MFLOP IEEE FPU, four 32-bit parallel DSP devices, a transfer controller with up to 480 MB/s off-chip transfer rate, and a video controller. All the processors are coupled tightly through an on-chip crossbar that provides shared access to on-chip RAM. Some common features of both processors include

- Single-chip parallel MIMD DSP.
- Master processor.
  - o   32 bit RISC processor.
  - o   4 KB instruction cache.
  - o   4 KB data cache.
- Transfer controller.
  - o   64-bit data transfers.
  - o   Up to 480 MB/s transfer rate.
  - o   32 bit addressing.
  - o   Intelligent queuing and cycle prioritisation.
- 4 GB address space.
- 16.6 ns cycle time.

While the 'C80 possesses four parallel processors with 32-bit advanced DSP devices, 2 KB instruction cache and 8 KB data RAM per parallel processor the C82 possesses two parallel processors with 32-bit advanced DSP processors, 4 KB instruction cache, 4 KB parameter RAM, and 8 KB of data RAM per parallel processor. The 'C80 possesses 50 KB on-chip RAM and the 'C82 possesses 44 KB of on-chip RAM.

Vector Processors

As mentioned before, vector processors are designed to efficiently handle arithmetic operations on elements of arrays, called vectors. These machines are extensively used in various fields especially in scientific computing, where matrix and vector arithmetic are common. A number of companies including Control Data Corporation (CDC), CRAY, Intel, IBM, Fujitsu, NEC, and Hitachi are renowned in the field of vector computing. To cover all the vector processors developed so far by different companies is beyond the scope of this book. To provide a flavour of the evolution of vector processors a selected set is discussed below (Cray, 2002; Intel, 2002).

CDC vector processors. CDC announced two vector processors namely, the CDC STAR-100 and the TI ASC in 1972. These processors could be categorised as memory-memory vector machines. Both of them possessed high start-up overhead and worked on vectors ranging from several hundred to several thousand elements. The crossover between scalar and vector could be over 50 elements. Following these, CDC developed CYBER-205, which was marketed in 1981. The CYBER-205 was developed on the basis of the same basic architecture of the STAR, however, its performance was improved in almost all areas compared to STAR. Its vector unit was expanded with up to four vector pipelines, each with multiple functional units and a wide load/store pipe that provided multiple words per clock. Like CDC STAR, CYBER-205 was a memory-memory vector processor. To keep the hardware simple and support high-bandwidth requirements (up to two memory references per FLOP), these machines did not efficiently handle unit stride. While most loops have unit stride, a no unit stride loop had poor performance on these machines because memory-to-memory data movements were required to gather the non-adjacent vector elements.

CDC formulated a group called ETA to build a new vector supercomputer namely ETA-10 in the early 1980s. The ETA used low-temperature CMOS in a configuration with up to 10 processors. Each processor retained the memory-memory architecture based on the CYBER-205. The ETA-10 was delivered in the late 1980s. It was capable of 10 GFLOPS. ETA-10 achieved enormous peak performance in many areas except its scalar speed, which was not comparable. CDC, the first supercomputer vendor, closed ETA and left the supercomputer design business in 1989.

Cray's vector processors. The Cray Company was established in 1976 and to date has deployed several vector processors, including CRAY-1, CRAY X-MP, CRAY-2, CRAY-3, CRAY SX-6 and CRAY SV1. The company announced its first vector processor CRAY-1 with an 80 MHz clock cycle in 1976. CRAY-1 was designed for supercomputing with pipelined vector arithmetic units. It was then the fastest scalar machine. The CRAY-1was a vector-register based machine. In the case of this type of machine vector pipelines are filled from the vector element values currently in the vector registers. This reduces the time to fill the pipelines (the startup time) for vector arithmetic operations; the vector registers can even be filled while the pipelines are performing some other operation. The vector results may be put back into a vector register after the completion of the operation, or they may be piped directly into another pipeline for an additional vector operation (chaining). The CRAY-1 was the first machine to use such chaining. Vector processors employing chaining techniques cannot only reduce the startup time for each operation compared to memory-memory machines, but are also able to execute two floating-point operations at the same time, doubling the number of MFLOPS.

Each vector register on the CRAY-1 and on later CRAY vector processors contains 64 single precision elements. An exception is the CRAY C90 with vector registers containing 128 elements. Each single precision element or word contains 64 bits; this is the equivalent of double precision values on most other machines. The CRAY-1 and later CRAY vector processors have twelve different pipelines or functional units. These are of several types such as: vector pipelines for integer or logical operations on vectors, vector/scalar pipelines for floating-point operations using scalars or vectors, scalar pipelines for integer or logical operations on scalars and address pipelines.

The CRAY X-MP was first delivered in 1983 with an improved clock rate, better chaining support, and multiple memory pipelines. The MP portion of the name refers to multiprocessing. Although the architecture of CRAY X-MP is similar to that of the CRAY-1, but it is more supportive for overlapped operations along with multiple memory pipelines. The CRAY X-MP is a shared-memory multiprocessor with each CPU controlling its own set of vector processors. The CRAY X-MP is capable of two reads and two writes per clock cycle. This allows a faster load and store between the memory and the vector registers and prevents some pipeline delays. Chaining on this machine allows all three floating-point pipelines to operate simultaneously. The CRAY X-MP favoured a multiprocessor approach, first offering a two-processor version and later a four-processor machine.

The CRAY-2, a completely new design configurable with up to four processors, was introduced in 1985. It possesses a much faster clock than the X-MP and also much deeper pipelines. It does not support chaining and also has enormous memory latency, and has only one memory pipe per processor. Thus, the CRAY-2 is an improvement on the CRAY X-MP only for those problems that require large amounts of memory.

In 1988, CRAY Research introduced the CRAY Y-MP - a bigger and faster version of the X-MP. The Y-MP allows up to eight processors and lowers the cycle time to 6 ns. With a full complement of eight processors, the Y-MP is generally the fastest supercomputer. A later version, called the CRAY C90, was announced in 1990, allowing up to sixteen processors. The CRAY Y-MP/M90 was introduced in 1991. This model is more like the CRAY Y-MP than the CRAY C90 but provides the ability to handle larger memory addresses and has a much larger memory. Partly because of the additional memory, the CRAY Y-MP/M90 operates at about half the speed of the CRAY C90. Originally limited to eight processors, the CRAY Y-MP/M90 has since been extended to handle up to sixteen processors. The CRAY Y-MP models are the fastest of the CRAY models discussed so far. The clock cycle time of both the CRAY Y-MP and the CRAY Y-MP/M90 is 6 nsec. The CRAY C90 has a clock cycle of 4.2 ns.

In 1989, a new company called CRAY Computer Corporation and headed by Seymour CRAY split from CRAY Research, Inc. Each company continued to develop supercomputers using the CRAY name. The CRAY Computer Corporation followed the ideas derived from the CRAY-2. The group also continued the work that begun before the split to develop the CRAY-3 (based on the CRAY-2). The work continued for about ten years, and the machine was never sold commercially. In 1993, one CRAY-3 was placed in the supercomputing centre at NCAR for research purposes. It contains four processors and has a clock cycle time of 2.1 ns.

In 2001, CRAY Inc. signed an agreement with NEC Corporation to market NEC's SX line of vector supercomputers worldwide alongside the company's current CRAY SV1™ and future CRAY SV2™ vector systems. The new CRAY SX-6, announced in October of 2001, is now available for customer shipment. The CRAY SX-6 series is among the world's most powerful supercomputers for challenging industrial, academic and civilian government research applications.

The CRAY SX-6 architecture uses parallel vector processor (PVP) nodes as building blocks. These building blocks provide elements of a symmetric multiprocessing (SMP) system. The CRAY SX-6 series also offers scalability to multiple-node configurations.

A single CRAY SX-6 node offers two to eight vector processors, each with 8 GFLOPS of peak performance. Each processor shares a uniform high-performance memory of 16 to 64 GB capacity, and up to 256 GB per second of bandwidth. Configurations of multiple CRAY SX-6 nodes, called MultiNodes, deliver a single-system image through an internode crossbar IXS. MultiNodes offer up to 8 TFLOPS (one trillion calculations per second) of peak performance, 8 TB of memory and an I/O bandwidth of more than 800 GB/s.

The CRAY SV1ex supercomputer combines single-processor performance with advances in custom architectural designs including the use of fifth-generation CMOS and industry-standard memory and disk technologies. The CRAY SV1ex system upholds the ability to run vector computations through high-bandwidth, vector cache memory making use of CMOS and SDRAM implementations.

Intel i860 vector processor. The Intel i860 (N10) vector processor delivers supercomputer performance in a single VLSI component. The 64-bit design of the i860 balances integer, floating point, and graphic performance. The i860 is used as the data processor in Intel's massively parallel Touchstone and Paragon supercomputers.

The i860's parallel architecture supports up to three operations per clock, one integer or control instruction per clock and up to two floating-point results per clock. It possesses a high level of integration on one chip containing 32-bit integer and control units, 32/64-bit pipelined floating point adder and multiplier units, 64-bit 3-D graphic unit, paging unit with translation look aside buffer, 32×32-bit integer register file, 16×16-bit FPU register file, 4 KB instruction cache, 8 KB data cache.

The i860XP (N11) is an extension to i860, with MP support (enable physical snooping), new process, and better performance. The i860XP microprocessor is designed around a RISC core and incorporates three additional functional units (multiply, add and graphics) within a pipelined floating-point unit. Available in 40 and 50 MHz versions, the four-way set associative 16 KB instruction and 16 KB data caches use a modified, exclusive, shared, or invalid (MESI) cache coherency protocol.

IBM/370 with IBM/3090 vector facility. IBM 3090 is a vector computer equipped with virtual storage addressing capability and vector registers. It resembles the vector facilities present in supercomputer such as the CRAY family. Three out of six of the IBM's 3090 central processors have vector facility, each with 128 KB of data cache, 256 MB of memory, and 256 MB of expanded memory. The IBM 3090 vector facility is shown in Figure 7.3 (IBM, 2002).

There are sixteen 32-bit vector registers in IBM 3090 (Stallings, 2003). The vector registers can also be coupled to form eight 64-bit registers. Any register can hold an integer or floating-point value. Thus, the registers can be used for 32-bit and 64-bit integer values as well as floating-point values. IBM 3090 vector organisation is referred to as register-to-register, because the vector operands can be staged in vector registers, an approach also used in CRAY super computers. The main advantage of this approach is that the operation is decoupled from slower main memory and instead takes place primarily with registers resulting in some speed enhancement. The 3090 can perform one main-storage access per processor.

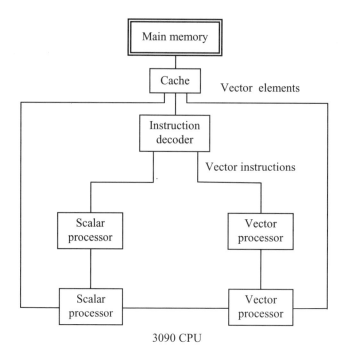

Figure 7.3. IBM 3090 with vector facility

7.5 A Comparative Overview

A significant number of microprocessors from various companies have been discussed in this chapter. Each of these microprocessors possess numerous features. To characterise a microprocessor precisely all these features require to be addressed, which is beyond the scope of this book. However, an attempt has been made to cover only the salient features. The era of the microprocessor began with the development of Intel 4004 4-bit microprocessor by the Intel Corporation. The current widespread use of microprocessors was beyond imagination at the early stages. However, soon after the release of 4004, Intel realised the prospects of microprocessors and engaged itself in the design and development of processors, and numerous other companies followed. As discussed, various architectural designs of general-purpose and special-purpose devices have emerged over the years. Among these the main architectures, namely CISC and RISC, are significant.

As discussed, CISC architecture is relatively older than RISC. The rationale of the CISC approach is that it requires fewer machine language instructions for a given high-level language program leading to shorter program execution times. This is true if complex instructions could be executed quickly and efficiently. However, this is quite challenging in practice. On the other hand, more RISC instructions are needed to perform a given task compared to its CISC counter part. However, RISC instructions are well suited to pipelining and they can be effectively used by optimising compilers (Hamacher *et al.*, 2002). There are some other advantages and disadvantages of both architectures. It seems that RISC architecture is gaining popularity day-by-day, and many companies are adopting RISC architecture to design their processors. Among the processors covered in this chapter, The ARM, PowerPC, Sun SPARC, HP's Alpha and Intel's IA-64 series are based on RISC architecture. On the other hand, Motorola 680X0 and Intel 80X86 processors are CISC architecture based.

7.6 Summary

Microprocessors constitute the main component of computer systems and as such have played a significant role in the development and activities of the modern world. They have added flexibility and convenience to the human life in many respects. These facilities and conveniences are being improved day by day with the improvement of microprocessors. To meet the multifarious needs of the people varied types of microprocessors have been developed and the use of microprocessors diversified at a rapid pace.

As sequential computing cannot cope with the speed demand of a number of applications, parallel processing has emerged as a potential solution, within which the microprocessor is the key component. With this view, the evolution, types and features of various microprocessors have been discussed in this chapter. The background to the development of a microprocessor includes some reasons, some objectives and there are some means to achieve those objectives on the basis of existing technology and expertise. In describing various microprocessors, efforts have been made to address those issues. Evolution of both general-purpose and special-purpose microprocessors has been addressed, with examples of series of microprocessors developed by some of the leading companies in the world. These include general-purpose microprocessors from Intel, Motorola, Apple-IBM-Motorola, DSP devices from Texas instruments, and vector processors from CDC, CRAY, Intel and IBM. However, there are many other processors from many other companies, which have not been discussed as they are beyond the scope of this book.

7.7 Exercises

1. What is a microprocessor? What are the fundamental components of a microprocessor?

2. Describe briefly the various functional units of a microprocessor.

3. Distinguish between general-purpose and special-purpose microprocessors.

4. What is meant by a CISC processor? Describe some of the characteristics of a CISC processor.

5. What is meant by a RISC processor? Mention some of the characteristics of a RISC processor.

6. Distinguish between CISC and RISC microprocessors.

7. Indicate the purpose and general characteristics of a DSP device?

8. What are the advantages of a DSP device over a general-purpose processor?

9. What is meant by a vector processor? What are the general features of a vector processor?

10. What is meant by pipelining? Describe the important features of pipeline processors.

11. Describe the important features of Intel 8086, 80286, 80386 and 80486 microprocessors.

12. Describe important characteristics of Intel Pentium, Pentium Pro and Pentium II microprocessors.

13. Why did Intel introduce Pentium III and Pentium 4 microprocessors? Describe their salient features.

14. Distinguish between Motorola 68040 and 68060 microprocessors.

15. Why did IBM, Apple and Motorola make an alliance to develop PowerPC microprocessors?

16. Describe some of the features of PowerPC 620 and MPC7450.

17. Mention a few series of DSP devices developed by Texas Instruments.

18. Describe salient features of the TMS320C40 DSP device.

19. Distinguish between the TMS320-67X floating-point DSP device and the TMS320C8X multimedia video processor.

20. Mention various features of the vector processors developed by CRAY .

21. Describe Intel's i860 processor.

22. Describe a few important features of the IBM 3090 vector facility.

8. Real-time Performances: Algorithms vs Architectures

8.1 Objectives

- To evaluate real-time performance through case studies.
- To explore the matching and mismatching of algorithms with architectures.
- To identify the characteristics of architectures based on their real-time performance.

8.2 Introduction

The performance demands in modern engineering applications have motivated the utilisation of complex and computationally intensive algorithms, including adaptive and intelligent control methodologies. This in turn has led to hard constraints on computing requirements of the hardware, to achieve the required sample times. Therefore, real-time performance in control applications where the use of advanced control methods is warranted becomes difficult to accomplish. Many demanding complex control processes cannot be satisfactorily realised with conventional uniprocessor and multiprocessor systems. Alternative strategies where multiprocessor-based systems are employed, utilising high-performance processors and parallel processing techniques, could provide suitable methodologies (Jones, 1989; Tokhi *et al.*, 1992).

In a conventional parallel system all the PEs are identical. This architecture can be described as homogeneous. However, many signal processing and control algorithms are heterogeneous, as they usually have varying computational requirements. The implementation of an algorithm on a homogeneous architecture is constraining and can lead to inefficiencies because of the mismatch between hardware requirements and hardware resources. In contrast, a heterogeneous architecture having PEs of different types and features can provide a closer match with the varying hardware requirements and, thus, lead to performance enhancement. However, the relationship between algorithms and heterogeneous architectures for real-time signal processing and control is not clearly understood. The mapping of algorithms onto heterogeneous architectures is, therefore, especially challenging. To exploit the heterogeneous nature of the hardware it is required to identify the heterogeneity of the algorithm so that a close match is forged with the hardware resources available (Baxter *et al.*, 1994).

One of the challenging aspects of parallel processing, compared to sequential processing, is how to distribute the computational load across the PEs. This requires consideration of a number of issues, including the choice of algorithm, the choice of processing topology, the relative computation and communication capabilities of the processor array and partitioning the algorithm into tasks and the scheduling of these tasks. It is essential to note that in implementing an algorithm on a parallel computation platform, consideration of

i.   the interconnection scheme issues;
ii. the scheduling and mapping of the algorithm on the architecture; and
iii. the mechanism for detecting parallelism and partitioning the algorithm into modules or sub-tasks;

will lead to computational speedup (Agrawal *et al.*, 1986; Crummey *et al.*, 1994).

This chapter presents the real-time processing requirements of the eight different algorithms described in Chapter 6. These were implemented on the heterogeneous and homogeneous parallel processing architectures, described Appendix B, on the basis of computing methods, described in Chapter 6. The uni-processor architectures investigated, include, an i860 processor, a SUN SPARCstation of TMS390S10 processor type, a T8 transputer processor and a TMS320C40 DSP processor. In contrast, the multiprocessor architectures, include, a homogeneous network of T8s, a heterogeneous architecture of an i860 and a T8, a homogeneous architecture of C40 DSPs and a heterogeneous network comprising a C40 and a T8. All the algorithms are implemented on the four uniprocessor architectures and the four different parallel architectures. The flexible beam and flexible manipulator simulation algorithms were further considered for implementation on more than two processor-based architectures. The 3L Parallel C and ANSI C compilers are used to investigate the performance of the architectures. Finally, comparative real-time computational aspects, including, execution time, speedup and efficiency are looked at to reveal the matching and mismatching of the algorithms with the architectures.

8.3 Adaptive Active Vibration Control

As discussed in the previous chapters, the adaptive AVC algorithm consists of three different algorithms, namely, a simulation algorithm, an identification algorithm and a control algorithm. These were implemented on a number of computing platforms. Results of these implementations are presented and discussed with the real-time performance of the algorithms and architectures in this section.

8.3.1 Beam Simulation Algorithm

To investigate the real-time implementation of the simulation algorithm and study the behaviour of the system, an aluminium type cantilever beam of length $L = 0.635\,\mathrm{m}$, mass $m = 0.037\,\mathrm{kg}$ and $\mu = 1.351$ was considered. It was found through numerical simulations that for the purpose of this investigation reasonable accuracy in representing the first few (dominant) modes of vibration is achieved by dividing the beam into 19 segments ($n = 19$). Therefore, $n = 19$ was chosen throughout the investigations presented here. Moreover, a sample period of $\Delta t = 0.3\,\mathrm{ms}$, which is sufficient to cover all the resonance modes of vibration of the beam, was selected.

The algorithm was implemented as a sequential process on uniprocessor based architectures. In the case of the multiprocessor based architectures, on the other hand, the algorithm was partitioned for equal load distribution among the processors. As discussed earlier in Chapter 6, the simulation algorithm can be implemented in fine-grain. For implementation on two-processor based architectures, the beam segments were distributed among the processors according to the performance of the PEs. In this manner, the total execution times achieved by the architectures, in implementing the simulation algorithm over 20,000 iterations, were measured. Figure 8.1 shows the execution times with uniprocessor and multiprocessor architectures, where, the required real time is calculated as the product of the sampling time $\Delta t = 0.3\,\mathrm{ms}$ and total number of iterations. It is noted that in this case the superscalar i860 RISC processor performed the fastest and the transputer T8 the slowest of the processors used. Table 8.1 shows the performance of the computing platforms used relative to that of the i860. The simulation algorithm, as discussed in Chapter 6, is mainly of a matrix based computational type for which the powerful vector processing resources of the i860 are exploited and utilised to achieve the shortest execution time. The C40 does not have such vector processing resources making it about six times slower than the i860. This implies that the C40 is not performing well in situations where the algorithm is matrix type and extensive runtime memory management is involved. In the case of the C40+C40, although the load on each PE has reduced to half the total load, due to the nature of the algorithm the C40s have not achieved a performance better than a single i860 vector processor. A single transputer, on the other hand, performed about 9.864 times slower than the i860, as the transputer does not have vector processing resources or built-in cache. This is further evidenced in the homogeneous network of two T8s which has performed nearly six times slower than a single i860. This is at a similar level to that achieved with a single C40. The main factor causing the C40+T8 network to perform slower than the C40+C40 network is the serial to parallel communication link utilised and incorporation of the slower T8 processor in this architecture. Due to the shared memory communication overhead and incorporation of the slower (T8) processor the i860+T8 network, has performed slower than a single i860 processor. The performance of the i860+T8 compared to the C40+T8 and the C40+C40 architectures is, however, significantly better. The high-performance SPARC with its RISC features achieved better performance compared to C40 and T8 based architectures. It is noted in Figure 8.1 that all the architectures utilised have achieved the real-time performance required.

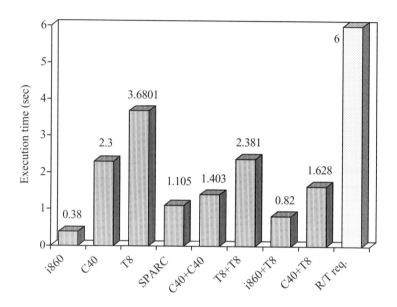

Figure 8.1. Execution times to implement the simulation algorithm on the computing platforms

Table 8.1. Execution times of various computing platforms *(T1)* relative to the vector processor i860 *(T2)* in implementing the simulation algorithm

	C40	T8	SPARC	C40+C40	T8+T8	i860+T8	C40+T8
T1/T2	6.052	9.684	2.907	3.692	6.266	2.158	4.284

To investigate the computation performance within fine-grain architecture, the algorithm was subdivided into fine-grain as discussed in Chapter 3. Figure 8.2 shows the real-time performance of the fine-grain transputer architecture. It follows from Figure 8.2 that, with increasing number of PEs, the computing performance has not increased linearly. This is due to communication overheads. It is noted that the performance remains nearly at a steady (saturation) level for over six PEs. It is further noted that the increase in communication overhead is more pronounced for less than five PEs and reaches a steady level for over five PEs. This is due to additional communication overhead between the PEs, which occur in parallel to others. The corresponding speedup and efficiency of the execution time for the network of T8s was discussed in Chapter 3.

Figure 8.3 shows the execution time of networks of C40s and T8s comprising three processors in implementing the algorithm. It is noted that compared to the network of T8s, the network of C40s achieved better performance. However, due to a mismatch between the nature of the algorithm with the architecture, communication overhead and runtime memory management problem, the network of C40s has not achieved an outstanding speedup and efficiency.

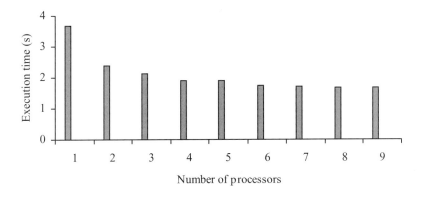

Figure 8.2. Execution time of the simulation algorithm on a transputer network

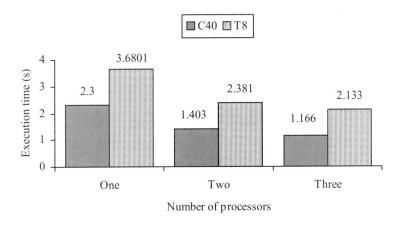

Figure 8.3. Execution time of the network of C40s and T8 for three processors when implementing the beam simulation algorithm

8.3.2 Identification Algorithm

As discussed in the previous chapter, the identification algorithm is composed of two components of similar nature and length, while estimating parameters of Q_0

and Q_1, and a process of controller design calculation. Thus, in the case of uni-processor based architectures, the algorithm was implemented as a sequential process. In the case of two-processor based architectures, on the other hand, the algorithm was partitioned so that the load at estimating parameters of Q_0 and Q_1 was equally distributed among the two PEs, with one of the PEs additionally carrying out calculation of controller parameters. Moreover, in this process, limited communication, due to the parameters of Q_0 and Q_1 for calculation of controller parameters, is required between the two PEs implementing the algorithm. Figure 8.4 shows the total execution times of the architectures to implement the identification algorithm over 1000 iterations. It is noted that among the architectures used the two C40s performed as the fastest and the T8 as the slowest. The execution times of the processors in relation to that of the i860 vector processor are given in Table 8.2. Although, the C40 was found to be faster than all the uni-processor based architectures, the speedup achieved in utilising two C40s compared to one C40 is only 1.35. This could be due to the nature of the identification algorithm, for which the pipelining nature of the C40 DSP device does not lead to a significant enhancement, even after reduction of the program into two segments.

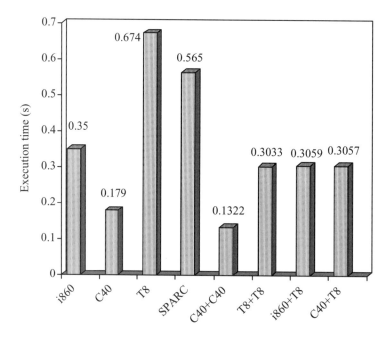

Figure 8.4. Execution times of the computing platforms to implement the identification algorithm

Table 8.2. Execution times of the computing platforms *(T1)* relative to that of the i860 (*T2*) to implement the identification algorithm

	C40	T8	SPARC	C40+C40	T8+T8	i860+T8	C40+T8
T1/T2	0.511	1.926	1.614	0.378	0.867	0.874	0.873

The algorithm does incorporate some matrix manipulation. However, as a result of the irregular nature of the algorithm, the i860 was found to perform even slower than the C40. In contrast, the T8 performed very well; the speedup achieved in using two T8s compared to one T8 is found from Table 8.2 to be 2.22. This super-linear speedup results from significant reduction in data handling for which the available internal memory of the T8 is sufficient, thus, reducing run-time memory management load. The performances in the case of the i860+T8 and the C40+T8 heterogeneous architectures were similar to one another and slower than a single C40. This is due to the involvement of the T8, which is slower than the i860 or the C40. The SPARC performed about 1.6 times slower than the i860 processor.

8.3.3 Control Algorithm

The control algorithm, as outlined and discussed in Chapter 6, is essentially composed of the simulation algorithm and the realisation of an IIR filter structure. Thus, in the case of uniprocessor based architectures the algorithm was implemented as a sequential process. In the case of two-processor based architectures, on the other hand, the algorithm was partitioned for the beam simulation part in a similar manner as in the case of the simulation algorithm with one of the PEs additionally carrying out calculation of the control signal. Figure 8.5 shows the execution times of the architectures used to implement the control algorithm over 20,000 iterations. It is noted that among the processors the i860 performed as the fastest and the T8 as the slowest. The execution times of the processors relative to that of the i860 RISC processor are shown in Table 8.3. Note that beam simulation forms a large proportion of the control algorithm. This makes the algorithm mainly an RI type. Thus, as in the case of the simulation algorithm, the powerful vector processing resources of the i860 are utilised to achieve the shortest execution time among the processors. The speedup achieved in using two C40s compared to one C40 is 1.6. The speedup achieved in using two T8s compared to one T8, on the other hand, is 1.54. These are consistent with the performance of these architectures in the case of the simulation algorithm. Thus, similar interpretations can be made on the speedup with the control algorithm. The performances of the i860+T8 and C40+T8 heterogeneous architectures are mainly attributed to the i860 and the C40 respectively.

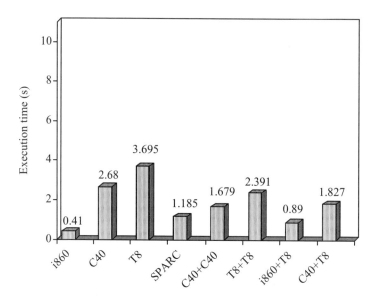

Figure 8.5. Execution times of the computing platforms to implement the control algorithm

Table 8.3. Execution times of the computing platforms *(T1)* relative to that of the i860 *(T2)* to implement the control algorithm

	C40	T8	SPARC	C40+C40	T8+T8	i860+T8	C40+T8
T1/T2	6.54	9.0	2.89	4.10	5.83	2.17	4.46

8.4 Adaptive Filtering

To investigate the performance of the architectures in realtime, implementation of adaptive filtering algorithms, a non-recursive filter realisation structure of 40 weights was used with the LMS algorithm and a second-order recursive filter realisation structure was used with the RLS algorithm. In both cases a block of 1000 samples input signal contaminated with random noise was used. These were implemented on uniprocessor and multiprocessor based heterogeneous and homogeneous architectures and their real-time implementation aspects studied, as described below.

8.4.1 LMS Filter

To investigate the real-time implementation of the LMS algorithm, the parameter $\mu = 0.04$ was used. The execution times achieved by the architectures, in implementing the LMS algorithm, are shown in Figure 8.6. It is noted that in this case the C40+C40 architecture performed the fastest and the T8 the slowest of the computing domains used. Table 8.4 shows the performance of the computing platforms used relative to that of the C40. The LMS algorithm, as noted earlier, incorporates some degree of irregularity due to the associated loops. The involvement of the regular DSP operations in the process, on the other hand, provides some degree of regularity in the algorithm as well. The exploitation of these two aspects of the process is evident in the performance of the C40 DSP device and the T8. A super-linear speedup of about 2.17 has been achieved with two C40s compared to a single C40. This is mainly due to a reduction of load on individual processors for which the internal cache of the C40 is sufficient, thus, not requiring extra memory and hence runtime memory management. A homogeneous architecture of two T8s, on the other hand, achieved a speedup of about 1.7 compared to a single T8. An influential factor in achieving lower speedup in the case of the T8+T8 architecture compared to the C40+C40 architecture is the serial communication link utilised in the T8s. The i860 is a vector processor and does not have the resources necessary to be exploited in implementing the LMS algorithm and, thus, performed 2.833 times slower than the C40, slightly faster than a single T8 and slower than the SPARC processor. The i860+T8 architecture achieved a longer execution time than a single i860 due to communication overheads, whereas in the case of the C40+T8 architecture both the C40 and the T8 contribute at a similar level to the performance achieved. Note that in the case of the C40+T8 architecture the serial to parallel communication link between the processors is very influential in slowing down the process.

8.4.2 RLS Filter

The execution times achieved by the architectures to implement the RLS filter algorithm are shown in Figure 8.7. It is noted that among the uniprocessors the C40 performed the fastest and the T8 performed the slowest. Among the multiprocessors, on the other hand, the C40+C40 performed the fastest and the T8+T8 the slowest. Table 8.5 shows the execution times of the processors relative to that of the C40 DSP device. The RLS algorithm is similar to the LMS algorithm except for the extra matrix computational components to evaluate the parameter vector and the covariance matrix. It is noted that, although the C40 performed the fastest among the processors in both uniprocessor and multiprocessor architectures, the speedup achieved with the C40+C40 architecture compared to a single C40 was 1.215. In the case of the T8s the speedup with T8+T8 compared with a single T8 was 1.102, which is not much different from the speedup achieved with the C40s. This is due to matrix computation as well as the increased level of data communication involved in the RLS filter algorithm. The i860 performed about

1.323 times slower than the C40. However, compared to the performances in the case of the LMS filter algorithm, the performance of the i860 is enhanced, mainly, due to the vector processing resources of the device. It follows from this result that the performance with the i860+T8 compared to a single T8 can mainly be attributed to the i860.

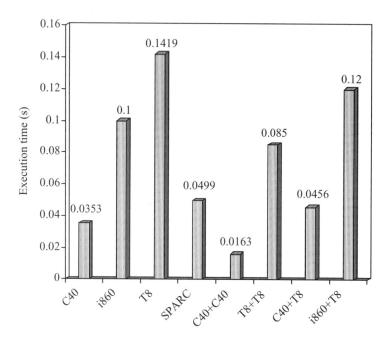

Figure 8.6. Execution times of the computing platforms to implement the LMS filter

Table 8.4. Execution times of the computing platforms *(T1)* relative to that of the C40 *(T2)* to implement the LMS filter

	i860	T8	SPARC	C40+C40	T8+T8	C40+T8	i860+T8
T1/T2	2.83	4.02	1.41	0.462	2.428	1.292	3.399

8.4.3 Relative Performances of the Architectures

Comparative measures of the performances of the uniprocessor and multiprocessor architectures in implementing the LMS and RLS filter algorithms are shown in Figures 8.8 and 8.9 respectively. It is noted that the C40 and C40+C40 architectures performed the fastest among the uniprocessor and multiprocessor architectures respectively to implement the LMS as well as the RLS filter algorithms. This can mainly be attributed to the high signal processing power and high-speed parallel interprocessor communication link incorporated within this processor. The SPARC performed faster than the i860 in the case of the LMS filter algorithm. In contrast, the i860 performed better than the SPARC for the RLS filter algorithm. Here, the vector processing resources of the i860 have been exploited to achieve better performance in the case of the RLS filter algorithm. These results are found to be consistent among the heterogeneous architectures where the C40+T8 performed the fastest for the LMS algorithm, whereas the i860+T8 performed the fastest for the RLS algorithm.

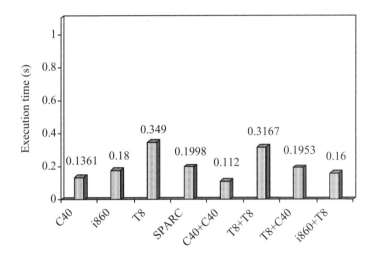

Figure 8.7. Execution times of the computing platforms to implement the RLS filter

Table 8.5. Execution times of the computing platforms *(T1)* relative to that of the C40 *(T2)* to implement the RLS filter

	i860	T8	SPARC	C40+C40	T8+T8	C40+T8	i860+T8
T1/T2	1.323	2.564	1.468	0.823	2.327	1.435	1.176

8.5 Spectral Analysis

The two most popular signal processing algorithms used for spectral analysis, namely, correlation and FFT, are considered here for real-time implementation. These algorithms were implemented on uniprocessor and multiprocessor based heterogeneous and homogeneous architectures and the results of these implementations were studied, as described below.

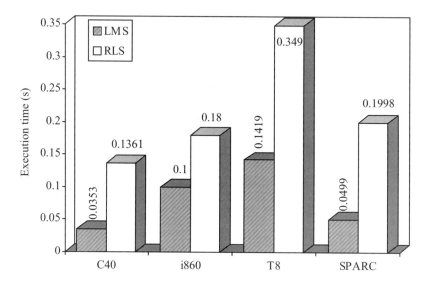

Figure 8.8. Comparative performance of the uniprocessor architectures to implement the LMS and RLS filters

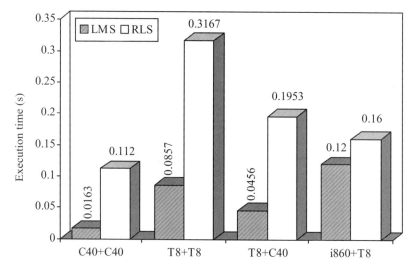

Figure 8.9. Comparative performance of homogeneous and heterogeneous architectures to implement the LMS and RLS filters

8.5.1 Correlation

To investigate the real-time implementation of the correlation algorithm, two waveforms, each of 1000 samples, were used. Figure 8.10 shows the real-time performance of the computing platforms when implementing the correlation algorithm. It is noted that the i860+T8 heterogeneous architecture performed the fastest and the T8 the slowest of the computing platforms. The i860 processor, on the other hand, achieved the shortest execution time among the uniprocessor platforms. In contrast, the T8 and C40 performed about 5.8 and 2.8 times slower than the i860, respectively. The SPARC processor achieved a longer execution time than the i860. The C40+T8 architecture performed faster than a single C40 but slower than the C40+C40 architecture. The T8 performed the slowest of all, possibly due to inefficient built-in RAM and lower processing power of the processor. The speedup achieved with the C40+C40 compared to a single C40 is 2.39. This super-linear speedup is mainly due to a significant reduction of load on each C40 to a level that requires less runtime data management and is suitable for the limited internal cache of the device. The speedup achieved with T8+T8 compared to a single T8 was only 1.25. This is due to the regular DSP nature of the algorithm, which is not as suitable for the T8 as for the C40. Moreover, the serial communication link incorporated with the T8+T8 affects the speed of the process significantly.

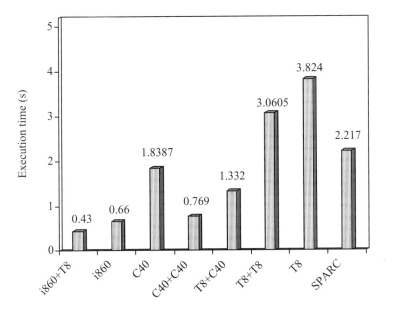

Figure 8.10. Execution times of the computing platforms to implement the correlation algorithm

8.5.2 FFT

The FFT algorithm is a regular and highly matrix based algorithm. Figure 8.11 shows the real-time performance of the computing platforms when implementing 512- and 256-point FFT algorithms. The i860, with its powerful vector processing resources, achieved the best performance among all the platforms. The i860+T8 heterogeneous architecture, on the other hand, achieved the shortest execution time among the parallel architectures. The C40 performed 10 to 15 times slower than the i860 when implementing the 256- and 512-point FFT algorithms respectively. In contrast, the SPARC processor achieved better performance in comparison to the C40 and T8 processor. The C40+T8 achieved a longer execution time than a single C40. Note that the FFT algorithm is highly regular and matrix based, thus not suitable for T8, which performed more than 30 times slower than the i860 vector processor.

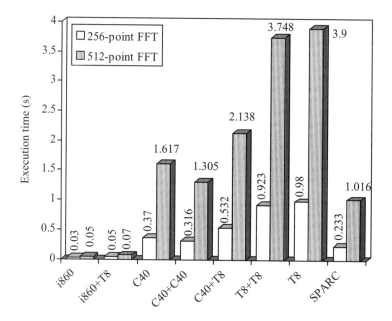

Figure 8.11. Execution times of the computing platforms to implement the FFT algorithm

Table 8.6. Execution time to implement a 512-point FFT (*T1*) over that to implement a 256-point FFT (*T2*) on the computing platforms

	i860	i860+T8	SPARC	C40	T8+T8	T8	C40+T8	C40+C40
T1/T2	1.67	1.400	4.36	4.37	4.062	3.98	4.019	4.135

The relative execution times of the computing platforms when implementing the 256-point and 512-point FFT algorithms are shown in Table 8.6. It is noted that the speedup when implementing the 256-point FFT algorithm compared to the 512-point FFT algorithm is not linear for all the computing domains. The i860 and the i860+T8 performed 1.7 and 1.4 times slower respectively to implement the 512-point FFT compared to implementing the 256-point FFT. All the other computing domains performed nearly four times slower when implementing the 512-point as compared to implementing the 256-point FFT algorithm. This implies that large amounts of data handling, i.e., run-time memory management and regular matrix based calculation requirements are highly detrimental to the performance of all the computing domains used except the i860.

8.5.3 Relative Performances of the Architectures

Figure 8.12 shows the comparative performance of the uniprocessor architectures when implementing the correlation and FFT algorithms. Note that the relative speedup with the FFT computation in comparison to the correlation algorithm is not the same for all the architectures. For instance, the i860 took 13.2 times longer with correlation than with the FFT. In contrast, the C40 required 1.14 times more execution time. This disparity implies that the FFT algorithm is implemented more efficiently than the correlation algorithm on the i860.

A comparison of the real-time performance of the parallel architectures is shown in Figure 8.13. It is noted that the i860+T8 heterogeneous architecture achieved the best real-time performance for both the correlation and FFT algorithms. This is due to the highly regular and vector type signal processing nature of the algorithms, which match the capabilities of the i860. The C40+C40 homogenous architecture achieved the second best performance for both algorithms. The T8+T8 homogenous architecture, on the other hand, achieved the longest execution time among the parallel architectures for both algorithms.

8.6 Flexible Manipulator System

The flexible manipulator simulation algorithm, as described in Chapter 6, was implemented as a sequential process on uniprocessor based architectures. For multiprocessor based architectures, on the other hand, the algorithm was partitioned for equal load distribution among the processors on the basis of the performance of the PEs. The real-time performances of different uniprocessor and multiprocessor computing domains are shown in Figures 8.14 and 8.15 respectively, where the general-purpose 386 and 486 are also shown for reasons of comparison. These are presented in ascending order of performance of the computing platforms for better presentation. The required real-time requirements indicate the actual time needed to implement the algorithm. It is noted that in this case the super-scalar i860 RISC processor performed the fastest and the T8 the slowest of the processors used.

Table 8.7 shows the performance of the computing platforms relative to that of the i860.

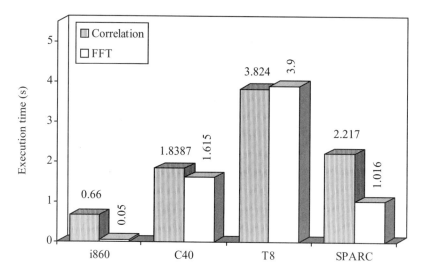

Figure 8.12. Comparative performance of uniprocessor based architectures when implementing the correlation and the 512-point FFT algorithms

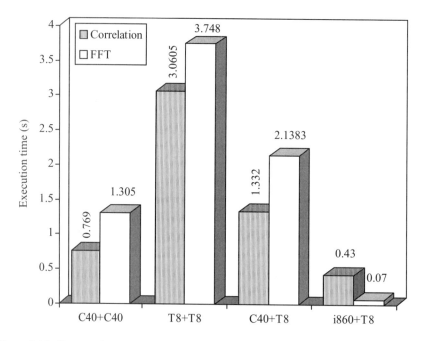

Figure 8.13. Comparative performance of homogeneous and heterogeneous architectures when implementing the correlation and the 512-point FFT algorithms

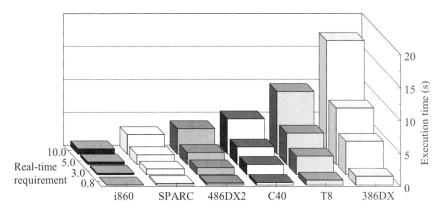

Figure 8.14. Execution time of uni-processor architectures when implementing the flexible manipulator simulation algorithm

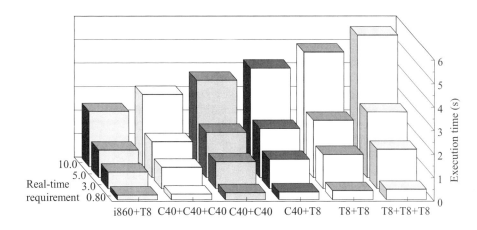

Figure 8.15. Execution times of the parallel architectures whein implementing the flexible manipulator simulation algorithm

Table 8.7. Execution times of the computing platforms *(T1)* relative to that of the i860 *(T2)* when implementing the flexible manipulator simulation algorithm

	C40	T8	SPARC	C40+C40	T8+T8	i860+T8	C40+T8
T1/T2	6.31	7.28	3.563	4.539	5.74	2.784	4.929

The flexible manipulator simulation algorithm, as discussed earlier, is mainly of a matrix based computational type for which the powerful vector processing resources of the i860 are exploited and utilised to achieve the shortest execution time among the processors. The C40 does not have such vector processing resources making it about 6.31 times slower than the i860. This implies that the C40 is not performing well in a situation where the algorithm is of matrix type and extensive runtime memory management is involved. A single transputer, on the other hand, performed about 7.28 times slower than the i860. The main factor causing the C40+T8 network to perform slightly slower than the C40+C40 network is the serial to parallel communication link utilised and the slower computation performance of the T8 in this architecture. Due to the shared memory communication overhead and incorporation of the slower (T8) processor, the i860+T8 network achieved a longer execution time than a single i860 processor. The performance of the i860+T8 compared to the C40+T8 and the C40+C40 architectures was significantly better. The SPARC performed slightly faster due to its RISC features.

Figure 8.16 shows a comparison of the execution times of the T8s and C40s homogeneous architectures. It is noted that the network of C40s achieved better performance than the network of T8s. Due to an increase in the communication overheads the performance with three T8s was lower than that with two T8s. In contrast, three C40s achieved slightly better performance than two C40s. This is due to the high-speed parallel communication links of the network of C40s. This is further demonstrated in Table 8.8 in terms of speedup and efficiency achieved with the networks. The speedups achieved with two C40s and three C40s compared to a single C40 were only 1.38 and 1.52. Although, with two and three C40s the load on a PE is significantly reduced due to the nature of the algorithm and communication overheads, the C40s did not achieve better performance than a single i860 vector processor. Note that the speedup with three T8s compared to a single T8 was 1.11, which is less than that with two T8s. This implies that an increase in the number of PEs leads to an increase in the communication overheads and hence is detrimental to real-time performance of the architecture when implementing this algorithm.

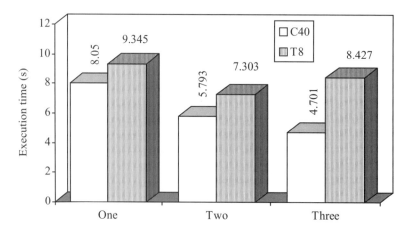

Figure 8.16. Comparative execution time of multiple C40s and T8s when implementing the flexible manipulator simulation algorithm

Table 8.8. Comparative speedup and efficiency with C40s and T8s

	Speedup		Efficiency	
Number of PEs	Two	Three	Two	Three
C40	1.38	1.52	69%	50%
T8	1.28	1.11	64%	37%

8.7 Comparative Performances of the Architectures

To explore the nature of the algorithms and architectures a comparative performance of the architectures in implementing the algorithms considered is discussed and presented in this section.

8.7.1 Uniprocessor Architectures

To investigate and compare the characteristics of the microprocessors for a number of different algorithms, the three most dominant uniprocessor architectures (i860, C40 and SPARC) are considered in one group. Comparative performances of the processors when implementing the algorithms are shown in Figure 8.17. For better presentation the execution times for the identification and LMS algorithms have been magnified two and eight times respectively. Moreover, the RLS and flexible manipulator simulation algorithms are similar to those of the identification and simulation algorithm respectively. Therefore, these two algorithms have not been considered here to avoid repetitive features of algorithms. It is noted that the i860 processor achieved the best performance among the processors for regular and vector based algorithms. In contrast, the C40 achieved the best performance with irregular algorithms. Compared to the C40, the SPARC processor achieved shorter execution times for regular and vector based algorithms but longer execution times for irregular type algorithms. The SPARC processor achieved better performance than the i860 processor only in the case of the LMS algorithm.

8.7.2 Multiprocessor Architectures

This section explores comparative characteristics of the heterogeneous and homogeneous architectures in relation to the algorithms. Figure 8.18 shows the execution times of the architectures in implementing the algorithms. For better presentation the execution times for the identification and LMS algorithms have been magnified two and eight times respectively. It is noted that the i860+T8 architecture achieved the best performance for beam simulation, beam control, FFT, correlation and the flexible manipulator simulation algorithms. In contrast, the C40+C40 architecture has achieved the best performance for the identification and

LMS algorithms. As discussed earlier and as further revealed here the i860+T8 heterogeneous architecture is more suitable for matrix based regular algorithms. In contrast, the C40+C40 homogeneous architecture is more suitable for irregular algorithms.

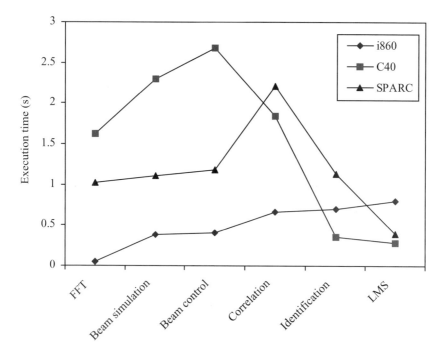

Figure 8.17. Execution times of the processors when implementing the algorithms

8.8 Summary

An investigation into the real-time implementation of several algorithms on a number of different computing domains has been presented. The real-time computing demand of complex and demanding algorithms for signal processing and control applications has been explored. Eight different computing platforms, including four homogeneous and heterogeneous parallel processing architectures have been investigated. The interprocessor communication speed of multiprocessor architectures has been investigated and presented. A comparative performance evaluation of each algorithm for different computing domains has been carried out and discussed on the basis of real-time computing aspects. Finally, comparative performance evaluations of algorithms and architectures have been carried out to demonstrate the matching and mismatching of algorithms and architectures.

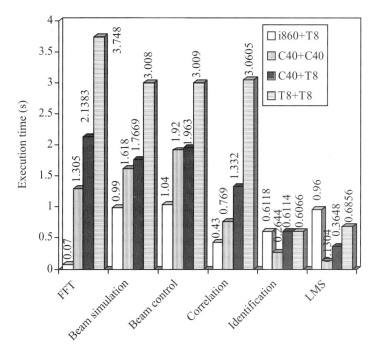

Figure 8.18. Execution times of the architectures when implementing the algorithms

8.9 Exercises

1. What are the key issues one has to consider in real-time parallel implementation? Among the architectural and software issues, which one is more important? Why?

2. Consider two microprocessors possessing the same specification and characteristics. Would the execution time of an algorithm in each of the processors be exactly same? If not, explain why?

3. In Figure 8.2, execution times of the simulation algorithm are at a similar level for 6, 7, 8 and 9 processors. Indicate the main reasons for this.

4. Implement the RLS and LMS filter algorithms using the same compiler on the same computing domain. Compare the performances in terms of execution time for the same input samples.

5. Based on the illustration provided in the case study, explain the impact of regularity and irregularity of an algorithm on performance of a processor

implementing the algorithm, if the processor is a vector processor or a DSP device.

Appendix A. List of Abbreviations

3D	Three-dimensional
ADE	Advanced dynamic execution
ALU	Arithmetic and logic unit
ARM	Advanced RISC machines
AVC	Active vibration control
ASB	Advanced system buffering
ATC	Advanced transfer cache
BIU	Bus interface unit
BPU	Branch processing unit
C40	Texas Instruments TMS320C40 DSP device
CAD	Computer-aided design
CC-NUMA	Cache-coherent NUMA
CDC	Control Data Corporation
CISC	Complex instruction set computer
CMOS	Complementary metal oxide semiconductor
CPI	Cycles per instruction
CPU	Central processing unit
CU	Control unit
DEC	Digital Equipment Corporation
DFT	Discrete Fourier transform
DIB	Dual independent bus
DMA	Direct memory access
DSP	Digital signal processing
ECC	Error correcting code
EPIC	Explicitly parallel instruction computing
EPROM	Erasable programmable read only memory
ETC	Execution trace cache
FD	Finite difference
FFT	Fast Fourier transform
FIR	Finite-impulse response
FPAU	Floating-point arithmetic unit

FPR	Floating-point register
FPU	Floating-point unit
FXU	Fixed-point unit
GB	Giga byte
GFLOPS	Giga floating-point operations per second
Gn	nth generation
GPR	General-purpose register
i860	Intel 80i860
IC	Integrated circuit
IIR	Infinite-impulse response
I/O	input/output
KB	Kilo byte
Kb	Kilo bit
L1	Level 1
L2	Level 2
L3	Level 3
LMS	Least mean squares
MAC	Multiply-accumulate
MB	Mega byte
Mb	Mega bit
MCA	Machine check architecture
MESI	Modified, exclusive, shared, or invalid
MFLOPS	Million floating-point operations per second
MIMD	Multiple-instruction multiple-data
MIPS	Million instructions per second
MMACS	Million multiply-accumulate per second
MMU	Memory management unit
MOPS	Million operations per second
MPU	Memory protection unit
NAS	Numerical aerodynamic simulation
NUMA	Non-uniform memory access
OTP	One-time programmable
PC	Personal computer
PARKBENCH	Parallel kernels and benchmarks
PDE	Partial differential equation
PE	Processing element
PG	Portland Group
PID	Proportional, integral, derivative
POSI	Portable operating system interface
POWER	Performance optimisation with enhanced RISC
PVP	Parallel vector processor
RAM	Random access memory
REE	Rapid execution engine
RISC	Reduced instruction set computer
RLS	Recursive least squares
ROM	Read only memory
SEC	Single edge contact

SIMD	Single-instruction multiple-data
SISD	Single-instruction single-data
SISO	Single-input single-output
SMP	Symmetric multiprocessing
SPEC	Standard performance evaluation corporation
SPMD	Single-program multiple-data
SSE	Streaming SIMD extensions
T8	T805 INMOS transputer
TB	Tera byte
UMA	Uniform memory access
VEA	Virtual environment architecture
VLIW	Very long instruction word
VPU	Vector-processing unit

Appendix B. Hardware and Software Resources

The hardware and software resources utilised in case studies throughout this book are describe in this appendix.

B.1 Hardware Resources

Conventional processors such as the Intel 386, Intel 486, Intel Pentium, M68040, VAX/8600 and IBM 390 fall into the family known as CISC architecture. Typical clock rates of today's CISC processors range from several hundred hertz to several gigahertz. Some modern CISC CPUs use partially hardwired control instead of microprogrammed control. The Intel i860, Texas Instrument microSPARC TMS390S10, MIPS R3000 and IBM RS/6000, on the other hand, fall into the RISC category. With the use of hardwired control, the cycles per instruction (CPI) of most RISC instructions has been reduced to 1 or 2 cycles. RISC processors possess limited addressing modes, a large number of general purpose registers and mostly split data cache and instruction cache. A special subclass of RISC processors is the superscalar processors, which allow multiple instructions to be issued simultaneously during each cycle. Thus the effective CPI of a superscalar processor should be lower than that of a generic scalar RISC processor. The clock rate of superscalar processors matches that of scalar RISC processors. The very long instruction word (VLIW) architecture uses even more functional units than a superscalar processor. Thus, the CPI of a VLIW processor can be further lowered. Due to the use of very long instructions, VLIW processors have been mostly implemented with microprogrammed control. Thus, the clock rate is slow with the use of read-only memory (ROM). A large number of microcode access cycles may be needed for some instructions (Hwang, 1993). In contrast, an important goal in DSP hardware design is to optimise both the hardware architecture and the instruction set for DSP operations. This is achieved by making extensive use of the concepts of parallelism. In particular, the key architectural features used are

    i.   Harvard architecture;
    ii. pipelining;
    iii. fast, dedicated hardware multiplier/accumulator;

iv. special instructions dedicated to DSP devices;
v. replication (more than one ALU, multiplier or memory unit);
vi. on-chip memory/cache.

The main advantage of DSP devices over general-purpose microprocessors is that they contain dedicated circuitry which provides high-resolution and high-speed arithmetic operations (Maguire, 1991).

Every microprocessor possesses its special features for specific applications. This leads to inherent difficulties in exploring comparative performances of different microprocessors. To explore the real-time performance in particular applications it is essential to implement the algorithm of that application into the PEs. This will also suggest static mapping of algorithms into PEs and performance evaluation of parallel architectures. The architectural features of PEs used in case studies throughout this book are given below.

B.1.1 Processing Elements

SPARC TMS390S10
SPARC is an acronym for Scalable Processor ARChitecture. Despite its independence from hardware implementation, the "scalable" part of the SPARC name refers to chip technology; specifically to the size of the smallest lines on the chip. The simple design of SPARC enables the chip design rules to be tightened easily (making the lines smaller) as fabrication technology improves (Hwang, 1993). The result is a chip with finer details and more compact layout that enables faster operation. The TMS390S10 is a RISC processor possessing individual floating-point unit, integer unit and memory management unit (MMU) with 50 MHz clock speed, on-chip data and instruction cache. This is a processor within the multitasking SUN SPARCstation, for which the performance at any time depends on the number of users. The most common features of the processor are as follows:

- SPARC™ compatible.
- High feature integration on-chip.
- Version 8 32-bit integer unit.
- Low-cost FPU, co-developed by Sun and Meiko, Ltd.
- Separate data and instruction caches.
- Complete integer unit.
- Single-issue, five-stage pipeline.
- Harvard architecture.
- 120 registers total (seven register windows).
- Peak execution rate of one cycle per instruction.
- Zero cycle branch penalty.
- Single-cycle loads and double-cycle stores (32 bits).
- Integer multiply and divide in hardware.
- FPU tailored for low cost and low power.
- Complete single and double precision IEEE floating-point math.
- Four cycle add and nine cycle multiply (double precision).

- Includes 32 32-bit registers.
- Separate on-chip caches.
- 4 KB of instruction cache: 32-bit wide access, direct mapped, physically tagged, virtually addressed, 28 lines of 32 bytes each.
- 2 KB of data cache: 32-bit wide access, direct mapped, physically tagged, 128.
- Lines of 16 bytes each. Supports a write-through, no write-allocate cache protocol.
- 32 entry, fully associative TLB.
- Supports up to 128 MB of system memory in four banks.
- 64-bit wide data bus to main memory.
- Economical single parity bit per 32-bit word.
- Supports 4 Mb and 16 Mb DRAM devices.
- Peak memory bandwidth is 115 MB/s.
- 25 MHz SBus clock frequency at 50 MHz.
- 32-bit data path.
- 50 MHz clock speed.
- Greater than 40 MIPs performance.
- 20-23 SPECint '92; 15-18 SPECfp '92.
- 288 pin, 10 mil outer lead bond (OLB) TAB packaging.
- 800,000 transistors.
- TI's 0.8 micron EPIC G1ZSE technology.

Intel i860

As discussed in Chapter 7, the i860 is a high-performance 64-bit vector processor with 40 MHz clock speed, a peak integer performance of 40 MIPS, 8KBytes data cache and 4KBytes instruction cache and is capable of 80 million floating-point operations per second (MFLOPS) for single precision and 60 MFLOPS for double precision. It is Intel's first superscalar RISC processor, possessing separate integer unit, floating-point unit, graphics unit, adder unit, multiplier unit and MMU. The i860 executes 82 instructions, including 42 RISC integer, 24 floating-point, 10 graphics, and 6 assembler pseudo operations in one clock cycle. All external or internal address buses are 32 bits wide, and the external data path or internal data bus is 64 bits wide. However, the internal RISC integer ALU is only 32 bits wide. The instruction cache transfers 64 bits per clock cycle, equivalent to 320 MB/s at 40 MHz. In contrast, the data cache transfers 128 bits per clock cycle. There are two floating-point units, namely, the multiplier unit and the adder unit, which can be used separately or simultaneously under the coordination of the floating point control unit. Special dual-operation floating-point instructions such as add-and-multiply and subtract-and-multiply use both the multiplier and adder units in parallel. Furthermore, both the integer unit and the floating-point control unit can execute concurrently (Hwang, 1993).

The TMS320C40 DSP Device

The TMS320C40 (C40) is a high-performance Texas Instruments 32-bit DSP processor with 40 MHz clock speed, 8KBytes on-chip RAM, 512 bytes on-chip instructions cache and is capable of 275 million operation per second (MOPS) and 40 MFLOPS. This DSP processor possesses six parallel high-speed communication links for interprocessor communication, 20 MB/s asynchronous transfer rate at each port for maximum data throughput, eleven operations/cycle throughput, resulting in massive computing and sustained CPU performance. In contrast, it possesses two identical external data and address buses supporting shared memory systems and high data rate, single-cycle transfers, separate internal program, data, and DMA co-processor buses for support of massive concurrent I/O of program and data throughput, thereby maximising sustained CPU performance (Texas Instruments, 1991d; Brown, 1991).

The T805 Transputer
The T805 (T8) is a general-purpose medium-grained 32-bit INMOS PP element with 25 MHz clock speed (yielding up to 20 MIPS performance), 4KBytes on-chip RAM and is capable of 4.3 MFLOPS. The T8 is a RISC processor possessing an on-board 64-bit floating-point unit and four serial communication links. The communication links operate at speeds of 20 Mb/s and achieve data rates of up to 1.7 MB/s unidirectionally or 2.3 MB/s bidirectionally. More importantly, the links allow a single transputer to be used as a node among any number of similar devices to form a powerful PP system. The transputer thus provides an important bridge between single-chip, real-time control and general-purpose real-time computer control systems, and, in effect, removes the current distinction between the two (Irwin and Fleming, 1992; Transtech Parallel Systems, 1991).

General-purpose Intel CISC Processors
A number of general-purpose processors were used to explore the real-time computational issues. This includes, Intel Pentium PIII (1000 MHz), Intel Pentium II (300 MHz), Intel Pentium (90 MHz), Intel 80486 and Intel 80386DX. Characteristics and most common features of these processors are discussed in Chapter 7.

B.1.2 Parallel Architectures

SIMD and MIMD are the two most widely used parallel architectures. The architectures are realised by interconnecting PEs to allow for interprocessor communication and synchronisation. Various interconnection topologies exist. To investigate the real-time implementation of the signal processing and control algorithms, four different types of SIMD/MIMD (depending on the application and mapping of the algorithm) are utilised. These are divided into two categories, namely, heterogeneous architecture and homogeneous architecture. As described earlier, an architecture comprising identical PEs is termed a homogeneous architecture. In contrast, when the PEs are different the architecture is termed heterogeneous. Two homogeneous and two heterogeneous architectures are considered in the book. These are briefly discussed below.

Homogeneous Architecture of T8s

A homogeneous network of T8s comprises of a Transtech TMB08 motherboard with 10 TRAMs possessing ten T8s. The root T8 incorporates 2MBytes local memory with the rest of the T8s each having 1MByte. The T8s are all identical 32-bit with 25 MHz clock speed as mentioned earlier. A SUN workstation is used as the host of the network for downloading the program into the network. The serial links of the processors are used to communicate with each other. Figure B.1 shows the topology of the transputer network. The topology shown is utilised as it is simple to realise. Moreover, it reflects the structure of the algorithms considered in this study.

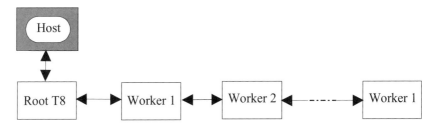

Figure B.1. Operational configuration of the homogeneous architecture incorporating a network of T8s

Heterogeneous Architecture of T8 and i860

The operational configuration of the heterogeneous parallel architecture which comprises an i860 RISC processor and a T8 is shown in Figure B.2. This includes an IBM compatible PC, A/D and D/A conversion facility, a TMB16 motherboard and a TTM110 board incorporating a T8 transputer and an i860 vector processor. The TTM110 board also possesses 16 MB shared memory and 4 MB of private memory accessible by the T8. The i860 and T8 processors communicate with each other via this 16 MB shared memory. The interface between the i860 and the shared memory is highly optimised to transport data at up to 160 MB/s. The T8 is used to boot the i860 by holding the i860 in reset while loading the memory with bootstrap code. Upon resetting the TRAM, the transputer is reset in the normal manner but the i860 is held in reset by the hardware, with HOLD asserted, releasing the bus to the transputer. The i860 code is then written to the boot address in the system RAM by the T8 before the i860 is released from reset by the transputer. The PC is used as a host for downloading the program into the T8 and i860 under control of the TMB16 motherboard (Transtech Parallel Systems, 1991).

Homogeneous Architecture of C40s

The homogeneous architecture of C40s comprises a Transtech match-box, a TMB08 motherboard and TDMB410 motherboard. A SUN workstation is used as host for downloading programs into the network. The TMB08 contains the root transputer T8 communicating through the Transtech match-box to the host. The root T8 possesses 1 MB local memory and communicates to the TDMB410 (C40 network) via a link adapter. The serial to parallel link is used for communication between the root T8 and the C40 network. On the other hand, the C40s communicate with each other via parallel links. Each C40 processor possesses 3 MB DRAM and 1 MB SRAM. The architectural features of the C40 processor were described earlier. Figure B.3 shows the topology of the network. The T8 in the network is used as the root processor providing an interface between the host and the first C40. The topology was chosen on the basis of the algorithm structure, which is simple and well reflected as a linear farm (Transtech Parallel Systems, 1993).

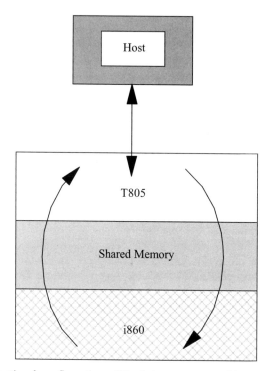

Figure B.2. Operational configuration of the heterogeneous architecture incorporating an i860 and a T8

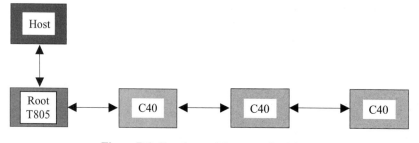

Figure B.3. Topology of the network of C40s

Heterogeneous Architecture of T8 and C40

This heterogeneous architecture integrates a T8 and C40, communicating with each other via serial to parallel or parallel to serial links. The operational configuration of the architecture is shown in Figure B.4. The T8 in the network is used both as the root processor providing interface with the host and as an active PE. This is basically, a part of the homogeneous network of C40s, where the root T8 processor and the first C40 are used as PEs of the heterogeneous architecture.

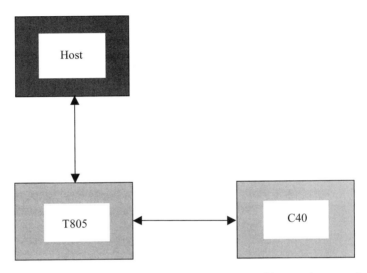

Figure B.4. Operational configuration of the heterogeneous architecture incorporating a C40 and T8

Dual-processor-based Multitasking Computing Domains

Two multiprocessor based multitasking computing domains were used to explore the real-time computing issues. One of these two domains comprises dual PII, 300MHz processors and the other one comprises dual PIII, 1000MHz processors. The manufacturer of the first one is HP and the second one is Dell. These are mostly shared memory based architectures and are not suitable for hard real-time signal processing and control system. These computing domains are used to explore comparative performances of the architectures. However, their use for multithread computing is very useful in exploring algorithm design issues for real-time applications. Most common features of these dual-processor based multi-tasking computing domain are available in the web site of the respective companies.

B.2 Software Support

Software support is needed for the development of efficient programs in high-level languages. The ideal performance of a computer system demands a perfect match between machine capability and program behaviour. Identifying the best compiler for the application in hand is, therefore, especially challenging due to the unpredictable runtime behaviour of compilers and the memory management capability. In signal processing and control applications it is important to select a suitable programming language that can support highly numerical computation for real-time implementation. The investigation here involves performance evaluation of some commonly used compilers on several computing platforms. The algorithms were coded in the high-level languages ANSI C, 3L Parallel C, Occam and Borland C as appropriate for the hardware used. Table B.1 shows the compilers with the corresponding computing platforms used.

Among these compilers, C/C++ compilers support almost all PP architectures. Between C/C++ and Occam, the Occam compiler is more straight-forward for PP, although it is not as efficient in numerical calculation as Parallel C or ANSI C (Bader and Gehrke, 1991). For implementations involving the T8, it is worth noting that the ANSI C compiler is used in investigations involving performance evaluations of hardware architectures when implementing the algorithms whereas 3L Parallel C and Occam are used in investigations involving performance evaluation of the compilers.

Table B.1. Compilers used for the different computing platforms

i860	C40	Single T8	SUN	386/486	Dual-Pentium II/III	Multiple T8	i860+T8
Portland Group ANSI C	3L Parallel C, V. 1.0.1	INMOS ANSI C, 3L Parallel C, V. 2.1 & Occam	ANSI C for UNIX	Borland C++	C/C++ Gcc	INMOS ANSI C & Occam	Portland Group ANSI C & INMOS ANSI C

B.3 Other Supporting Resources

The computing domains used in the case studies in this book involve several motherboards and slave boards, comprising different microprocessors, control module and interfacing module. These are the TMB16 and TMB08 motherboards, TTM110 board, Transtech MatchBox and TDMB410 board. Among these the TMB16/TMB08 has been used as the main motherboard. These are described below.

B.3.1 TMB Motherboard

TMB16/TMB08 are part of the TMB motherboard series, which, is a full length PC hosted transputer module (TRAM) motherboard, with space to plug in up to ten TRAMs. The module motherboard allows access to transputer modules from a variety of host machines. The motherboard can be divided into two distinct parts: the host-specific interface and the generic TRAM part. It can be used to:

i.   configure the transputer into any topology;
ii. build computing systems consisting of any mix of TRAMs;
iii. chain a number of motherboards together;
iv. run and test applications on transputers without first having to configure the links.

The architectures that evolved to meet these requirements incorporate the following important features:

- The modules in a network are connected in a pipeline using two links from each module.
- The remaining links from each module are available for user configuration either by direct wiring (via edge connectors) or via a programmable link switch (IMS C004). When using C004 switches:
  - a number of links are taken from the C004 to the edge connector
  - each C004 is controlled by a T212 (T2) transputer
  - the T2 transputers are connected together in a separate pipeline.
- The first module in the pipeline on the motherboard can control a subsystem of other modules that may reside on the same motherboard, another motherboard or be distributed across a number of boards.
- An interface may be provided to allow non-transputer based host computers to control and communicate with the TRAMs on the motherboard.

The motherboard has an advanced 16-bit interface to the host PC giving raw data transfer rates between the PC and the motherboard of up to 4.9 MB/s. In practice, all of this data has to pass down a single transputer link, which limits the available bandwidth (from PC to module0 link0) to around 1.25–1.5 MB/s. The motherboard can be hosted within computers with a PC AT BUS (IBM AT and all clones), with a VME bus (Sun), with a MCA Bus (IBM PS/2) and with the NuBus (Macintosh).

The Host Interface

The TRAM standard defines that TRAM motherboards have on them a T2 transputer to control the board's IMS C004 electronic link switch. In the TMB16, a portion of this transputer's external memory space is mapped directly to the PC. This allows the transputer to communicate with the PC's Intel microprocessor directly and at the high bandwidth supported by the PC's backplane. This design fully exploits the presence of the T2, which is in its own right, a powerful 16-bit processor.

Since the external memory interface of the T2 is 16 bits wide, the TMB16 interface transfers 16-bit quantities (words) at a time. The PC's backplane is normally clocked at 8 MHz. Given that it requires three clock cycles to transfer a word across the backplane, this gives a theoretical backplane bandwidth of (8/3) MWords/s or 5.33 MB/s. Measurements show that the TMB16 can transfer data across the backplane at 3.5 MB/s at this bus speed, realising in practice 70% of the available bandwidth. The interfacing techniques for TMB08 are similar to those for TMB16.

The Control Structure

In order to be able to program and use the TRAMs on the motherboard there is a mechanism for their control. The most important control signal sent to a TRAM is the reset signal. When a TRAM receives this signal the transputer is reset to an initial state. The transputer has to be in this state before a user program can be loaded onto it. The host computer controls all of the processors in the system. This configuration is suitable for 3L scientific languages and the Occam toolset, where the various tools in the programming environment (compilers, linkers, etc.) are invoked from the host computer. Figure B.5 shows the control architecture.

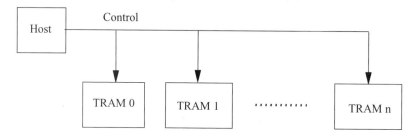

Figure B.5. Control architecture

Operation of Software

The operation of the interface is controlled by software running on the T2. The software is provided on the TMB16 in non-volatile memory (EPROM). Essentially, the software exploits the available hardware to provide an invisible connection between the PC and the transputers mounted on the motherboard. So as to maintain backwards compatibility with older motherboards, this connection is actually between the PC and link0 of TRAM module0. At the beginning of any data communication the two processors must agree on the amount of data to be transferred. This is achieved by transmitting a signal 16-bit counter. That counter determines the number of bytes that follow. Due to hardware limitations the maximum amount of data that can be transferred in one block is limited to 32KB.

As far as the T2 is concerned, access to the PC is achieved by reading/writing it to external memory. During a data transfer from the PC to the transputer network, T2 reads from its external memory and writes data to one of its links. This is achieved with the transputer channel out instruction. Similarly, during a data transfer from the transputer network to the PC, the T2 reads data from a link and writes this data to its external memory, which is achieved with the transputer channel in instruction.

Network Configuration

The interconnection between the module slots, the electronic link switch, the link patch area and the edge connector, is shown in Figure B.6. The connected links to the C004 are:

- Link0 of all TRAM slots except module0 (module0 link0 is connected to the host PC via the T2).
- Link3 of all TRAM slots.
- Eight edge connectors.
- Two spare links, which are taken to the patch area.

The connections made by the C004 are controlled by the T2 transputer acting in the role of a configuration processor. The T2 is, in turn, controlled via ConfigUp. Using the NCS program the C004 switch connection can be changed.

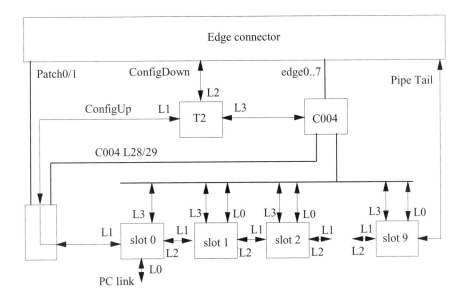

Figure B.6. Network configuration

B.3.2 The TTM110 Board

The TTM110 is an i860 based TRAM, which offers the advantages of high-performance vector processing to the transputer based parallel processing community. Intel's i860 64 bit RISC microprocessor is used to deliver up to 80 MFLOPS of vector based computational performance alongside a 32-bit T805 transputer. The TTM110 is ideal for parallel applications that are computationally intensive and require performance greater than that available from transputers. Application particularly suited include:

- Signal and image processing.

- Seismic exploration.
- Computational physics and chemistry.
- Finite element analysis.
- Computational fluid dynamics.
- Any other areas with significant levels of vector and/or matrix manipulation requirements.

The TTM110 comprises an Intel i860 processor clocked at 40 MHz, a T805 transputer clocked at 25 MHz , 4 or 16 MB of shared memory and 1 or 4 MB of private memory accessible only by the transputer. It could be interfaced to Sun, Silicon Graphics, DEC, PCs and IBM PS/2 host computers. The TTM110 communicates with other processing units using four 20 Mb bi-directional transputer links. It is also capable of controlling other TRAMs and transputer based systems using its subsystem port.

B.3.3 Transtech MatchBox

The Transtech MatchBox is designed to interface the SCSI port of a workstation to Transtech PARAstation systems. However, the MatchBox can equally be used to interface to a wide range of other INMOS transputer based system. On multi-user workstations, up to four users can concurrently access the PARAstation resources per MatchBox.

B.3.4 The TDMB410 Motherboard

The TDMB410 is a four-slot motherboard. The TDMB410 has both an 8 and a 16-bit host interface which can drive a single C40 communication link. The link from the interface can be connected either to the first on-board slot or to the boards edge connector. The motherboard is capable of sharing a maximum of 16 MB of memory between the slot sites via the C40's global memory busses. The memory is organised into four banks and supports multiprocessor access avoiding many memory contention difficulties.

References

Agrawal DP, Janakiram VK, Pathak GC (1986) Evaluating the performance of multicomputer configuration. IEEE Computer 19: 23–37

Akl SG (1989) The design and analysis of parallel algorithms. Prentice Hall, New York

Amdahl GM (1967) Validity of the single-processors approach to achieving large scale computing capabilities. Proceedings of AFIPS 30: 483–485

Anderson AJ (1991) A performance evaluation of microprocessors, DSPs and the transputer for recursive parameter estimation. Microprocessors and Microsystems 15: 131–136

ARM (2002) http//:www.arm.com (2002)

Asche RR (1996) Win 32 multithreading performance. URL: http://www.scis.nova.edu/~linje /cs770/win32.html, Nova Southeastern University, Florida

Azad AKM (1994) Analysis and design of control mechanisms for flexible manipulator systems, PhD Thesis. Department of Automatic Control and Systems Engineering, The University of Sheffield, UK

Bader G, Gehrke E (1991) On the performance of transputer networks for solving linear systems of equation. Parallel Computing 17: 1397–1407

Barragy E, Carey GF, Geijn RV (1994) Performance and scalability of finite element analysis for distributed parallel computation. Journal of Parallel and Distributed Computing 21: 202-212

Baxter MJ, Tokhi MO, Fleming PJ (1994) Parallelising algorithms to exploit heterogeneous architectures for real-time control systems. Proceedings of IEE Control-94 Conference, Coventry, 21–24 March, vol. 2, pp 1266–1271

Braunl T (1993) Parallel programming - an introduction. Prentice Hall International, UK

Bennett S (1988) Real-time computer control: An introduction. Prentice-Hall, UK

Brey BB. (2003). The Intel microprocessors. Pearson Education, New Jersey

Brown A (1991) DSP chip with no parallel? Electronics World+Wireless World October: 878–879

Carr AG (1990) Real-time signal processing - a review of applications and technology. Computing and Control Engineering Journal 1: 77–80

Chen CS (1982) Application of on-chip signal processor in digital controller implementation. The IEEE Control Systems Magazine 2 (9): 16–22

Ching PC, Wu SW (1989) Real-time digital signal processing system using a parallel architecture. Microprocessors and Microsystems 13(10): 653–658

Costa LDF (1989) Bit reversal for the TMS32010 fast Fourier transform calculation. Microprocessors and Microsystems 13 (7): 445–448

Cray (2002) http//:www.cray.com (1995 – 2002)

Crown CFN, Grant P M (1985) Adaptive filters. Prentice-Hall, Inc., USA

Crummey TP, Jones DJ, Fleming PJ, Marnane WP (1994) A hardware scheduler for parallel processing in control applications. Proceedings of IEE Control-94 Conference, Coventry, 21-24 March, vol. 2, pp 1098–1103

Culler DE, Sing JP, Gupta A (1999) Parallel computer architecture. Morgan Kaufmann Publishers Inc., San Francisco, California

Cvetanovic Z (1987) The effects of problem partitioning, allocation, and granularity on the performance of multiple processor systems. IEEE Transactions on Computers C-36 (4): 421–432

Denning PJ (1986) Parallel computing and its evolution. Communications of the ACM 29: 1163-1167

Fleming PJ (1988) Parallel processing in control - the transputer and other architectures, Peter Peregrinus, UK

Flynn MJ (1966) Very high-speed computing system. Proceedings of the IEEE 54: 12-27

Flynn MJ (1972) Some computer organisations and their effectiveness. IEEE Transactions on Computers 21 (9): 948-960

Grama AY, Gupta A, Kumar V (1993) Isoefficiency: Measuring the scalability of parallel algorithms and architectures. IEEE Parallel and Distributed Technology 1 (3): 12–21

Gray JS (1998) Interprocess communications in UNIX. Prentice Hall, USA

Gustafson JL (1988) Re-evaluating Amdahl's law. Communications of ACM 31 (5): 532–533

Gustafson JL (1992) The Consequences of fixed-time performance measurement. Proceedings of 25th International Conference on Systems Sciences, vol. III. IEEE Computer Society Press, Los Alamitos, California, pp. 113–124

Hamacher VC, Vranesic ZG, Zaky SG (2002) Computer organization. McGraw-Hill, London

Hays JP (1988) Computer architecture and organisation, McGraw-Hill, London

Hayes JP (2003) Computer architecture and organization. McGraw-Hill, London

Hocney RW, Jesshope CR (1981) Parallel computers. Hilger Publishing Co, Bristol

Hossain MA (1995) Digital signal processing and parallel processing for real-time adaptive noise and vibration control, PhD thesis. Department of Automatic Control and Systems Engineering, The University of Sheffield, UK

Hwang K (1993) Advanced computer architecture – parallelism scalability programmability. McGraw-Hill, California

Hwang K, Briggs FA (1985) Computer architectures and parallel processing. McGraw-Hill, California

IBM (2002). http//:www.ibm.com (2002)

Ifeachor EC, Jervis BW (1993) Digital signal processing - A practical approach. Addison–Wesley, UK

Intel (2002). http//:www.intel.com (2002)

Irwin GW, Fleming PJ (1992) Transputer in real-time control. John Wiley and Sons, England

Jones DI (1989) Parallel architecture for real-time control. Electronics and Communication Engineering Journal 1: 217–224

Kabir U, Hossain MA, Tokhi MO (2000) Reducing memory access time in real-time implementation of signal processing and control algorithms. Proceedings of AARTC00: IFAC Workshop on Algorithms and Architectures for Real-time Control, Palma de Mallorca, Spain, 15-17 May 2000, pp 15–18

Khokhar AA, Prasanna VK, Shaaban ME, Wang C (1993) Heterogeneous computing: challenges and opportunities. IEEE-Computer 26 (6): 18–27

King GA, Picton PD (1990) Vector signal processors and digital filters in data compression for electronic publishing. Microprocessors and Microsystems 14 (9): 555-563

Kourmoulis PK (1990) Parallel processing in the simulation and control of flexible beam structure system, PhD Thesis. Department of Automatic Control and Systems Engineering, The University of Sheffield, UK

Lapidus L (1982) Numerical solution of partial differential equations in science and engineering. John Wiley and Sons, New York

Lawes ST, Clarke T (1994) Real-time aircraft modelling and control using parallel processing. Proceedings of IEE Control-94 Conference, Coventry, 21–24 March, vol. 2, pp 1094–1097

Li Y and Rogers E (1991) Progress in the development of parallel adaptive/self-tuning controllers. Proceedings of IFAC Workshop on Algorithm and Architectures for Real-time Control, Bangor, 11–13 September, pp 207–211

Lo JL, Barroso LA, Eggers SJ, Gharachorloo K, Levy HM, Parekh SS (1998) An analysis of database workload performance on simultaneous multithreaded processors. Proceedings of 25th Annual International Symposium on Computer Architecture, USA, June 1998

Ludeman CL (1986) Fundamentals of digital signal processing. Harper & Row, USA

Madeira MM, Tokhi MO, Graca Ruano M (1998) Comparative study of sequential and parallel implementations of a doppler signal estimator, Research report no. 773. Department of Automatic Control and Systems Engineering, The University of Sheffield, UK

Maguire LP (1991) Parallel architecture for Kalman filtering and self-tuning control, PhD Thesis. The Queen's University of Belfast, UK

Megson GM (1992) Practical steps towards algorithmic engineering. IEE Digest No. 1992/204. Colloquium on Applications of Parallel and Distributed Processing in Automation and Control, London

Moldovan DI (1993) Parallel processing – from applications to systems. Morgan Kaufmann Publishers, USA

Motorola (2002). http//:www.motorola.com (August 2002)

Nation WG, Maciejewski AA, Siegel HJ (1993) A methodology for exploiting concurrency among independent tasks in partitionable parallel processing systems. Journal of Parallel and Distributed Computing 19: 271–278

Nichols B, Buttar D, Farrel JP (1996) Pthreads programming. O'Reilly & Associates, USA

Nocetti GDF, Fleming PJ (1991) Performance studies of parallel real-time controllers. Proceedings of IFAC workshop on Algorithms and Architectures for Real-time Control, Bangor, UK, 11–13 September 1991, pp 249–254

Nussbaum D, Agrawal A (1991) Scalability of parallel machines. Communications of the ACM 34 (3): 57–61

Portland Group Inc. (1991) PG tools user manual. Portland Group Inc., USA.

Proakis JG, Manolakis DG (1988) Introduction to digital signal processing. Macmillan Publishing Company, USA

Ramos-Hernandez DN (1998) Performance evaluation, mapping and scheduling for heterogeneous architectures, PhD thesis. Department of Automatic Control and Systems Engineering, The University of Sheffield, UK

Redstone JA, Eggers SJ, Levy HM (2000) An analysis of operating system behavior on a simultaneous multithreaded architecture. Proceedings of the 9th International Conference on Architectural Support for Programming Languages and Operating Systems, USA, November 2000

Shaheed, M.H. (2000). Neural and genetic modelling, control and real-time finite element simulation of flexible manipulators, PhD thesis. Department of Automatic Control and Systems Engineering, The University of Sheffield, UK

Silverschatz A, Galvin PB (1998) Operating system concepts. Addison-Wesly Longman Inc., USA

Sing JP, Hennessy JL, Gupta A (1993) Scaling parallel programs for multiprocessors: methodology and examples. IEEE-Computer 26 (7): 42-50

Stallings W (2003) Computer organization and architecture. Pearson Education, New Jersey

Stankovic JA (1997) Real-time and embedded systems. In: Tucker, A.B. Jr (ed.) The computer science and engineering handbook. CRC Press Inc., USA

Stone HS (1990) High performance computer architecture. Addison-Wesley, USA

Sun (1994) http//:www.sun.com (1994 – 2002)

Sun X-H, Gustafson J (1991) Toward a better parallel performance metric. Parallel Computing, 17 (10): 1093–1109

Sun X-H, Ni L (1993) Scalable problems and memory-bounded speedup. Journal of Parallel and Distributed Computing, 19 (September): 27–37

Sun X-H, Rover DT (1994) Scalability of parallel algorithm-machine combinations. IEEE Transactions on Parallel and Distributed Systems 5 (6): 599–613

Tan M, Siegel HJ (1998) A stochastic model for heterogeneous computing and its application in data relocation scheme development. IEEE Transactions on Parallel and Distributed Systems 9 (11): 1088-1101

Texas Instruments (1991a) TMS320C40 User's Guide. Texas Instruments, USA

Texas Instruments (1991b) TMS320C4x User's Guide. Texas Instruments, USA

Texas Instruments (1991c) TMS320 floating-point DSP optimising C compiler User's Guide. Texas Instruments, USA

Texas Instruments (1991d) TMS320 User's Guide. Texas Instruments, USA

Texas Instruments (2002) http//:www.ti.com (1995 – 2002)

Thoeni UA (1994) Programming real-time multicomputers for signal processing. Prentice-Hall, Hertfordshire

Tokhi MO, Hossain MA (1994) Self-tuning active vibration control in flexible beam structures. Proceedings of IMechE-I: Journal of Systems and Control Engineering 208 (I 4): 263-277

Tokhi MO, Hossain MA (1995) CISC, RISC and DSP processors in real-time signal processing and control. Microprocessors and Microsystems 19 (5): 291–300

Tokhi MO, Hossain MA (1996) Real-time active control using sequential and parallel processing methods. In: Crocker MJ, Ivanov NI (eds) Proceedings of the fourth International Congress on Sound and Vibration, St Petersburg, 24–27 June 1996, vol 1. International Scientific Publications, Auburn, USA, pp 391–398

Tokhi MO, Leitch RR (1992) Active noise control. Oxford Science Publications, Clarendon Press, Oxford

Tokhi MO, Virk GS, Hossain MA (1992) Integrated DSP3 systems for adaptive active control. Colloquium on Active Techniques for Vibration Control - Sources, Isolation and Damping, London, 28 October 1992, IEE Digest No. 1992/185, IEE, UK, pp 6/1–6/4

Tokhi MO, Hossain MA, Baxter MJ, Fleming PJ (1995) Heterogeneous and homogeneous parallel architectures for real-time active vibration control. IEE Proceedings-D: Control Theory and Applications 142 (6): 1–8

Tokhi MO, Chambers C, Hossain MA (1996) Performance evolution with DSP and transputer based systems in real-time signal processing and control applications. Proceedings of UKACC International Conference on Control-96, Exeter, 02–05 September 1996, vol 1, pp 371–375

Tokhi MO, Hossain MA, Chambers C (1997a) Performance evaluation of DSP and transputer based systems in sequential real-time applications. Microprocessors and Microsystems 21: 237-248

Tokhi MO, Hossain MA, Baxter MJ, Fleming PJ (1997b) Performance evaluation issues in real-time parallel signal processing and control. Journal of Parallel and Distributed Computing 42 (1): 67-74

Tokhi MO, Shaheed MH, Ramos-Hernandez DN, Poerwanto H (1999a) Finite element simulation of a flexible manipulator-part 1: Sequential processing techniques. Journal of Low Frequency Noise, Vibration and Active Control 18 (4): 191-205

Tokhi MO, Shaheed MH, Ramos-Hernandez DN, Poerwanto H (1999b) Finite element simulation of a flexible manipulator-part 2: Parallel processing techniques. Journal of Low Frequency Noise, Vibration and Active Control 18 (3): 149-165

Transtech Parallel Systems (1991) Transtech parallel technology. Transtech Parallel Systems Ltd., UK

Transtech Parallel Systems (1993) Transtech parallel technology. Transtech Parallel Systems Ltd., UK

Tullsen DM, Lo LJ, Eggers SJ, Levy HM (1999) Supporting fine-grained synchronization on a simultaneous multithreading processor. Proceedings of the 5th International Symposium on High Performance Computer Architecture, USA, January 1999

Virk GS, Kourmoulis PK (1988) On the simulation of systems governed by partial differential equations. Proceedings of IEE Control-88 Conference, pp 318–321

Widrow B, Glover JR, McCool JM, Kaunitz J, Williams CS, Hearn RH, Zeidler JR, Dong E, Goodlin RC (1975) Adaptive noise cancelling: principles and applications. Proceedings of the IEEE 63: 1692 - 1696

Worley PH (1990) The effect of time constraints on scaled speedup. SIAM Journal of Scientific and Statistical Computing 11 (5): 838–858

Yan Y, Zhang X, Song Y (1996) An effective and practical performance prediction model for parallel computing on nondedicated heterogeneous NOW. Journal of Parallel and Distributed Computing 38 (1): 63–80

Zhang X, Yan Y (1995) Modeling and characterizing parallel computing performance on heterogeneous networks of workstations. Proceedings of Seventh IEEE Symposium on Parallel and Distributed Processing, San Antonio, Texas, 25–28 October 1995, pp 25–34

Index

5440